The Open University
M381 Number Theory and Mathematical Logic
Mathematics and Computing
A third level course

Number Theory

Unit 8
Diophantine Equations

Prepared for the Course Team by Alan Best

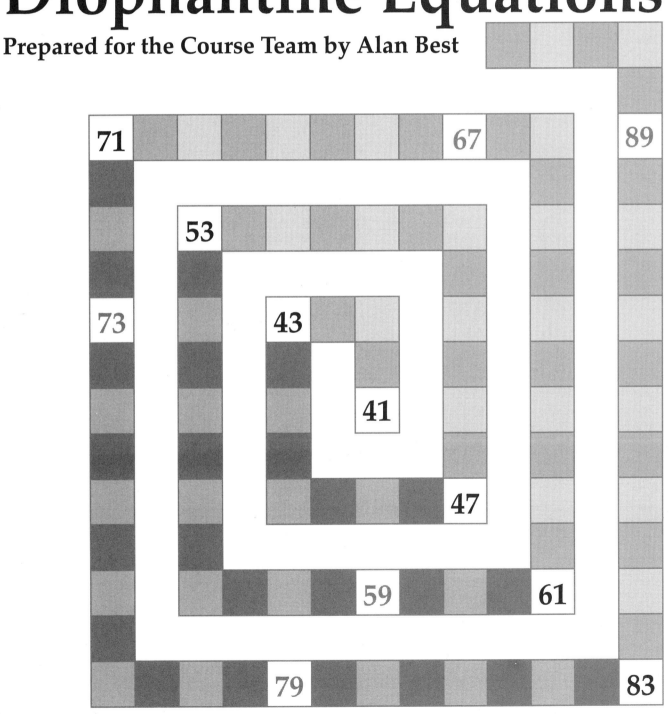

The M381 Number Theory Course Team

The Number Theory half of the course was produced by the following team:

Alan Best *Author*
Andrew Brown *Course Team Chair* and *Academic Editor*
Roberta Cheriyan *Course Manager*
Bob Coates *Critical Reader*
Dick Crabbe *Publishing Editor*
Janis Gilbert *Graphic Artist*
Derek Goldrei *Critical Reader*
Caroline Husher *Graphic Designer*
John Taylor *Graphic Artist*

with valuable assistance from:

CMPU *Mathematics and Computing, Course Materials Production Unit*
John Bayliss *Reader*
Elizabeth Best *Reader*
Jeremy Gray *History Reader*
Alison Neil *Reader*

The external assessor was:

Alex Wilkie *Reader in Mathematical Logic, University of Oxford*

The Open University, Walton Hall, Milton Keynes, MK7 6AA.

First published 1996. Reprinted 1997, 2001.

Edited, designed and typeset by the Open University using the Open University TEX System.

Printed and bound in the UK by The Charlesworth Group, Huddersfield

ISBN 0 7492 6449 7

This text forms part of an Open University Third Level Course. If you would like a copy of *Studying with The Open University*, please write to the Central Enquiry Service, PO Box 200, The Open University, Walton Hall, Milton Keynes, MK7 6YZ. If you have not already enrolled on the Course and would like to buy this or other Open University material, please write to Open University Educational Enterprises Ltd, 12 Cofferidge Close, Stony Stratford, Milton Keynes, MK11 1BY, United Kingdom.

CONTENTS

INTRODUCTION

A Diophantine equation is an equation in two or more variables which is to be solved within the set of integers. Diophantus was the first to consider such matters, his interest stemming from geometrical problems involving squares and cubes. Diophantus was only concerned with finding one solution rather than all solutions, and he was content with the solutions being rational numbers rather than integers. Nevertheless it is fitting that this immense branch of number theory is named after him.

Practically nothing is known of Diophantus the person. He lived in Alexandria around 250 AD and wrote in Greek, although there is a suggestion that he might have been of Babylonian origin. His fame rests on his *Arithmetica*; it is believed that this consisted of thirteen books but only six have survived. In addition to posing many of the famous problems in number theory which will occupy our attention in this unit, *Arithmetica* presents the first real treatise of algebra introducing, as it does, revolutionary mathematical notation and symbolism.

For fourteen or so centuries after Diophantus, little advance of any significance was made on general methods of solving Diophantine equations. Then Fermat arrived on the scene, his contributions in this area being regarded by many as the real origins of modern number theory.

Our first encounter with a Diophantine equation was in *Unit 1* where we considered the linear equation $ax + by = c$. Thinking geometrically this equation represents a line in the Cartesian plane. To solve it as a Diophantine equation requires finding all (if any) lattice points through which the line passes. There are a number of questions which we can ask, and have asked, of a linear Diophantine equation.

Recall that a lattice point is one with integer coordinates.

- Are there any integer solutions?

- Is the set of integer solutions finite?

- Is there a method of systematically finding all solutions?

We had little difficulty supplying answers to all three questions in the case of the linear Diophantine equation. But the same questions can be asked of any polynomial equation $P(x_1, x_2, \ldots, x_n) = 0$ in n variables and, more often than not, answering these questions presents considerable difficulty, even when $n = 2$ and P is a fairly simple polynomial. For example, consider the two variable equation $x^3 - y^2 = 2$. With a little trial and error you might be able to spot one solution and so answer the first of the above questions, but you would find the other two questions much more daunting. (Fermat managed them but we shall not go into his solutions here.)

In this unit we shall concentrate on a few of the more famous Diophantine equations. We begin where we left the previous unit, looking at applications of continued fractions.

1 PELL'S EQUATION

1.1 Pell's equation and continued fractions

Being unable to interest his contemporaries in his researches in number theory, Fermat took to issuing challenges to Europe's best mathematicians, with England's John Wallis being a prime target. Some of Fermat's unpublished discoveries came to light via these arithmetic challenges. One of his first posers was the following from 1657.

> Given a non-square positive integer n, find an integer y such that $ny^2 + 1$ is also a square. If a general rule cannot be discovered find the smallest values of y for the cases $n = 61$ and $n = 109$.

The underlying problem here is to solve the Diophantine equation $x^2 - ny^2 = 1$ for a general n, with the subsidiary challenge of finding particular solutions for the cases $n = 61$ and $n = 109$. (The trivial solution $x = 1$, $y = 0$ is discounted, the real task being to find solutions in positive integers.) Wallis together with his patron, Viscount Brouncker (who was the first president of the Royal Society), discovered a general method of solution, though they were unable to prove that their method always works. However the choice of posed special cases, $n = 61$ and 109, leaves no doubt that Fermat too was aware of some way of solving the problem. The values of y in the smallest solutions turn out to be $y = 226\,153\,980$ for $n = 61$ and $y = 15\,140\,424\,455\,100$ for $n = 109$ and these solutions were not likely to be found by trial and error! In contrast, for the adjacent values $n = 60, 62, 108$ and 110 the y values in the smallest solution turn out to be $y = 4, 8, 130$ and 2 respectively.

Fermat's posed problem was by no means the first appearance of the Diophantine equation $x^2 - ny^2 = 1$. The ancient Greeks had been considering the cases $n = 2$ and $n = 3$ in searching for rational approximations to $\sqrt{2}$ and $\sqrt{3}$ respectively. The equation for the case $n = 2$ also arises in the search for numbers which are both triangular and square. A famous problem posed by Archimedes, concerning breeds of cattle on the island of Sicily, was reduced by elementary algebra to the task of finding positive integer solutions of $x^2 - 4\,729\,494y^2 = 1$. The smallest solution turns out to have y as an integer of 41 digits and, not surprisingly, it was not known to the Greeks!

We saw how to find these in Unit 7.

We shall solve the triangular square problem in this section.

In 1759 Euler showed that any solution of $x^2 - ny^2 = 1$ must necessarily have $\dfrac{x}{y}$ as a convergent of $\sqrt{n}$ and he went on to discover a general method of solution based on the continued fraction of $\sqrt{n}$. His paper contained all that was needed to show that the Diophantine equation $x^2 - ny^2 = 1$ has infinitely many solutions and that all of them are obtainable from the continued fraction of $\sqrt{n}$, although he failed to collate them into a complete proof. Lagrange provided this in 1768.

The equation $x^2 - ny^2 = 1$ is known as *Pell's equation*. So where does Pell enter the story? It transpires that Euler wrongly attributed the first method of solution by Wallis and Brouncker to the English mathematician John Pell (1611–1685), and although this error is now recognized, the name of Pell has remained firmly attached to this Diophantine equation.

So much for the background, let us now make a start at solving this equation. If n is a non-square positive integer then $\sqrt{n}$ is irrational and so the equation $x^2 - ny^2 = 0$ (or $\dfrac{x^2}{y^2} = n$) has no integer solutions. But if $\dfrac{x}{y}$ is a good rational approximation to $\sqrt{n}$ then $\dfrac{x^2}{y^2}$ is close to n, or what amounts to the same thing, $x^2 - ny^2 = k$ is a small (positive or negative) integer.

As k cannot equal 0 the next best thing is that $k = \pm 1$. Having seen that the convergents are the 'best' rational approximations to $\sqrt{n}$, it will therefore come as no surprise to discover that every positive solution of $x^2 - ny^2 = 1$ arises as a convergent $\dfrac{x}{y}$ of $\sqrt{n}$. The groundwork for proving this has been done in the previous unit.

If $x = a$, $y = b$ is a solution then, for example, $x = -a$, $y = b$ is another solution. We shall confine attention to positive solutions.

Theorem 1.1 Solutions of Pell's equation are convergents

If $x = a$, $y = b$ is a positive solution of $x^2 - ny^2 = 1$ then $\dfrac{a}{b}$ is a convergent of $\sqrt{n}$.

Proof of Theorem 1.1

Making use of Theorem 4.3 of *Unit 7*, we aim to show that if $a^2 - nb^2 = 1$ then

$$\left| \sqrt{n} - \frac{a}{b} \right| < \frac{1}{2b^2},$$

which is sufficient to ensure that $\dfrac{a}{b}$ is a convergent of $\sqrt{n}$.

As $a^2 - nb^2 = 1$ we have $(a - b\sqrt{n})(a + b\sqrt{n}) = 1$ and so,

$$a - b\sqrt{n} = \frac{1}{a + b\sqrt{n}} > 0 \quad \text{and} \quad a > b\sqrt{n}.$$

Therefore

$$\left| \sqrt{n} - \frac{a}{b} \right| = \frac{a - b\sqrt{n}}{b} = \frac{1}{b(a + b\sqrt{n})}$$

$$< \frac{1}{b(b\sqrt{n} + b\sqrt{n})} = \frac{1}{2b^2 \sqrt{n}} < \frac{1}{2b^2}.$$

Hence $\dfrac{a}{b}$ is a convergent of $\sqrt{n}$. ∎

Knowing that all the solutions of Pell's equation are to be found amongst the convergents of $\sqrt{n}$ it remains to identify which, if any, of the convergents give rise to solutions. In the example below we have taken the case $n = 7$ and worked out a few of the convergents $\dfrac{p_k}{q_k}$ of $\sqrt{7}$. We have then calculated the corresponding values of $p_k^2 - 7q_k^2$ in the hope that the value 1 might turn up.

Example 1.1

Determine the convergents of $\sqrt{7}$ up to C_{10} and find which of these give rise to a solution of $x^2 - 7y^2 = 1$.

The continued fraction of $\sqrt{7}$ is $[2, \langle 1, 1, 1, 4 \rangle]$. Hence, using the tabular method to determine the convergents, we have the following.

k	1	2	3	4	5	6	7	8	9	10
p_k	2	3	5	8	37	45	82	127	590	717
a_k	2	1	1	1	4	1	1	1	4	1
q_k	1	1	2	3	14	17	31	48	223	271
$p_k^2 - 7q_k^2$	-3	2	-3	1	-3	2	-3	1	-3	2

This reveals two solutions to this Pell's equation, namely

$$x = 8, \ y = 3 \quad \text{as} \quad 8^2 - 7 \times 3^2 = 1,$$
$$x = 127, \ y = 48 \quad \text{as} \quad 127^2 - 7 \times 48^2 = 1.$$

However there is also the beginnings of a clear pattern in the values for $p_k^2 - 7q_k^2$. It looks as if the sequence of values $-3, 2, -3, 1$ is cycling. Could it be that C_4 and C_8 and every fourth convergent thereafter will give a solution? ◆

Problem 1.1

(a) Determine the convergents of $\sqrt{3} = [1, \langle 1, 2 \rangle]$ as far as C_{10} and check which of these satisfy $p_k^2 - 3q_k^2 = 1$.

(b) In the same way, find three convergents of $\sqrt{10} = [3, \langle 6 \rangle]$ which give solutions of $x^2 - 10y^2 = 1$.

Any lingering hope that each convergent would give rise to a solution of the corresponding Pell's equation have now been dispelled. On the positive side, however, for $n = 3$, 7 and 10, the convergents have, in each case, led to at least one solution. There are also encouraging signs from the row of $p_k^2 - nq_k^2$ in each of the constructed tables; the values here, like the partial quotients in the ICF of $\sqrt{n}$, appear to cycle. But the key observation to make from these examples concerns where the solutions arise. For $n = 3$ and for $n = 7$ the solutions of Pell's equation arise from the convergents corresponding to the penultimate partial quotient in the cycle. For example, the cycle in the continued fraction of $\sqrt{7}$ ends in a 4 and it is the convergents calculated from the partial quotient 1 immediately prior to this 4 which give the solutions.

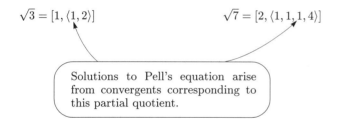

$$\sqrt{3} = [1, \langle 1, 2 \rangle] \qquad\qquad \sqrt{7} = [2, \langle 1, 1, 1, 4 \rangle]$$

Solutions to Pell's equation arise from convergents corresponding to this partial quotient.

Figure 1.1   Solving Pell's equation for the cases $n = 3$ and $n = 7$

The case $n = 10$ is similar, but Problem 1.1 part (b) gives more information. As the cycle in the ICF of $\sqrt{10}$ has length 1, every partial quotient comes immediately prior to the end of a cycle. But they do not all give rise to solutions; it appears that the even convergents give rise to a solution of the equation $x^2 - 10y^2 = 1$ while the odd convergents give rise to solutions of $x^2 - 10y^2 = -1$. This time it looks as if the selected convergents all give rise to solutions of $x^2 - 10y^2 = \pm 1$. We shall show that this is the case shortly.

1.2   Solution of Pell's equation

In the proof of the next theorem we need to make use of a result which we gave without proof in Section 3.3 of *Unit 7*. This result states that, in general, the ICF of $\sqrt{n}$ has the form

$$\sqrt{n} = [a_1, \langle a_2, a_3, \ldots a_3, a_2, 2a_1 \rangle].$$

Let the number of partial quotients in the cycle of $\sqrt{n}$ be s. Then, with the familiar notation $\dfrac{p_k}{q_k}$ for the convergents of $\sqrt{n}$,

We say that the *cycle length* of the ICF of $\sqrt{n}$ is s.

$$\frac{p_s}{q_s} = [a_1, a_2, a_3, \ldots a_3, a_2]; \qquad \frac{p_{2s}}{q_{2s}} = [a_1, a_2, a_3, \ldots a_3, a_2, 2a_1, a_2, a_3, \ldots a_3, a_2]$$

and, in general, $\dfrac{p_{rs}}{q_{rs}}$ is the convergent obtained by terminating in the rth cycle immediately before the final partial quotient $2a_1$. We are now in a position to state our main result.

Theorem 1.2 Solution of Pell's equation

If the continued fraction of $\sqrt{n}$ has cycle length s then

$$p_{rs}^2 - nq_{rs}^2 = (-1)^{rs}; \quad r = 1, 2, 3, \ldots,$$

and all solutions of

$$x^2 - ny^2 = \pm 1$$

are given in this way.

So, according to the theorem, as $\sqrt{14} = [3, \langle 1, 2, 1, 6 \rangle]$ has cycle length $s = 4$, the convergents $\dfrac{p_4}{q_4}, \dfrac{p_8}{q_8}, \dfrac{p_{12}}{q_{12}}, \ldots$ satisfy $p_k^2 - 14q_k^2 = 1$. On the other hand as $\sqrt{29} = [5, \langle 2, 1, 1, 2, 10 \rangle]$ has odd cycle length $s = 5$, the convergents

$$\frac{p_{10}}{q_{10}}, \frac{p_{20}}{q_{20}}, \frac{p_{30}}{q_{30}}, \ldots$$

satisfy $p_k^2 - 29q_k^2 = 1$, while the convergents

$$\frac{p_5}{q_5}, \frac{p_{15}}{q_{15}}, \frac{p_{25}}{q_{25}}, \ldots$$

satisfy $p_k^2 - 29q_k^2 = -1$.

In what follows we are going to prove only the main part of the theorem, namely that the listed convergents do indeed give solutions as claimed. We shall omit proof of the converse, namely that all solutions of $x^2 - ny^2 = \pm 1$ arise in this way. It is not difficult to prove that any solution of $x^2 - ny^2 = \pm 1$ must have $\dfrac{x}{y}$ as a convergent of $\sqrt{n}$. Rather than embark on a messy proof here that it must be one of the claimed convergents, we shall give, in Theorem 1.3, an alternative approach to obtaining 'all' the solutions.

Proof of Theorem 1.2

For each $r \geq 1$ we can write $\sqrt{n}$ as the non-simple finite continued fraction

$$\sqrt{n} = [a_1, a_2, a_3, \ldots a_3, a_2, 2a_1, a_2, a_3, \ldots, a_3, a_2, x]$$

with a total of $rs + 1$ partial quotients, the last of which, x, is not an integer. In fact

$$\begin{aligned} x &= [\langle 2a_1, a_2, a_3, \ldots, a_3, a_2 \rangle] \\ &= a_1 + [a_1, \langle a_2, a_3, \ldots, a_3, a_2, 2a_1 \rangle] \\ &= a_1 + \sqrt{n}. \end{aligned}$$

The final three convergents in the above finite continued fraction for $\sqrt{n}$ are

$$\frac{p_{rs-1}}{q_{rs-1}}, \quad \frac{p_{rs}}{q_{rs}} \quad \text{and} \quad \frac{xp_{rs} + p_{rs-1}}{xq_{rs} + q_{rs-1}}.$$

Now the last of these three is equal to $\sqrt{n}$ itself and so

$$\sqrt{n}(xq_{rs} + q_{rs-1}) = xp_{rs} + p_{rs-1},$$

and substituting $a_1 + \sqrt{n}$ for x, we get

$$\sqrt{n}((a_1 + \sqrt{n})q_{rs} + q_{rs-1}) = (a_1 + \sqrt{n})p_{rs} + p_{rs-1}.$$

This simplifies to

$$\sqrt{n}(a_1 q_{rs} + q_{rs-1} - p_{rs}) = a_1 p_{rs} + p_{rs-1} - nq_{rs}.$$

The right-hand side of this equation is an integer while the left-hand side is $\sqrt{n}$ times an integer. Since $\sqrt{n}$ is irrational, the only way that equality can occur is when both sides are equal to zero. Hence we have the two equations

$$a_1 q_{rs} + q_{rs-1} = p_{rs}$$

and

$$a_1 p_{rs} + p_{rs-1} = n q_{rs}.$$

Multiplying the first of these equations by p_{rs} and the second by q_{rs} and then subtracting gives

$$p_{rs}^2 - n q_{rs}^2 = p_{rs} q_{rs-1} - p_{rs-1} q_{rs},$$

the right-hand side of which is equal to $(-1)^{rs}$ by Theorem 1.3 property (a) of *Unit 7*. ∎

When the cycle length s of the continued fraction of $\sqrt{n}$ is even we have $(-1)^{rs} = 1$ and Theorem 1.2 tells us that $x = p_{rs}$, $y = q_{rs}$ is a solution of the Pell's equation for each $r \geq 1$. On the other hand, when s is odd, $x = p_{rs}$, $y = q_{rs}$ is a solution of $x^2 - ny^2 = -1$, when r is odd, and is a solution of $x^2 - ny^2 = 1$, when r is even.

Problem 1.2 _____

Given that $\sqrt{11} = [3, \langle 3, 6 \rangle]$, find the three smallest positive solutions of $x^2 - 11y^2 = 1$.

Problem 1.3 _____

Given that $\sqrt{13} = [3, \langle 1, 1, 1, 1, 6 \rangle]$, find one positive solution of each of the equations

$$x^2 - 13y^2 = 1 \quad \text{and} \quad x^2 - 13y^2 = -1.$$

To complete our survey of Pell's equation, we shall demonstrate an alternative way of finding all the solutions.

1.3 Solutions from the fundamental solution

We shall refer to the solution of $x^2 - ny^2 = 1$ in which x and y take their least positive values as the *fundamental solution*. From Theorem 1.2 we know that, if the ICF of $\sqrt{n}$ has cycle length s then the fundamental solution is given by $x = p_s$, $y = q_s$ when s is even, and by $x = p_{2s}$, $y = q_{2s}$ when s is odd.

There is a simple algorithm for constructing all solutions from the fundamental one.

If we have two solutions, $x = x_1$, $y = y_1$ and $x = x_2$, $y = y_2$, then we can readily deduce from the equation $x^2 - ny^2 = 1$ that $x_1 < x_2$ if, and only if, $y_1 < y_2$. So the solution with the smallest x value will also have the smallest y value.

Theorem 1.3 All solutions from a fundamental solution

Let $x = x_1$, $y = y_1$ be the fundamental solution of $x^2 - ny^2 = 1$. Then, for each integer $k \geq 1$, $x = x_k$, $y = y_k$ is also a solution, where the positive integers x_k and y_k are given by

$$x_k + y_k \sqrt{n} = \left(x_1 + y_1 \sqrt{n} \right)^k.$$

Conversely, the solutions given in this way are the only positive solutions.

Before embarking on the proof let us look at an example to be sure we understand the process implied in the statement of the theorem.

Example 1.2

Find four solutions of $x^2 - 8y^2 = 1$.

Since $\sqrt{8} = [2, \langle 1, 4 \rangle]$, which has a cycle of length 2, the second convergent $C_2 = \dfrac{3}{1}$ gives the fundamental solution $x_1 = 3$, $y_1 = 1$. Note that $3^2 - 8 \times 1^2 = 1$.

Further solutions are found as follows.

$$\left(3 + 1 \times \sqrt{8}\right)^2 = 17 + 6\sqrt{8},$$

so $x_2 = 17$, $y_2 = 6$ is a solution.

$$\left(3 + 1 \times \sqrt{8}\right)^3 = \left(17 + 6\sqrt{8}\right)\left(3 + \sqrt{8}\right) = 99 + 35\sqrt{8},$$

so $x_3 = 99$, $y_3 = 35$ is a solution.

$$\left(3 + 1 \times \sqrt{8}\right)^4 = \left(17 + 6\sqrt{8}\right)\left(17 + 6\sqrt{8}\right) = 577 + 204\sqrt{8},$$

so $x_4 = 577$, $y_4 = 204$ is a solution. ◆

Our proof of Theorem 1.3 is not particularly difficult but does involve a good deal of careful algebra. You will not be expected to reproduce this proof and can omit it if you so wish.

Proof of Theorem 1.3

The proof that each $x = x_k$, $y = y_k$ is a solution follows quickly from the observation that

$$\left(x_1 + y_1\sqrt{n}\right)^k = x_k + y_k\sqrt{n} \iff \left(x_1 - y_1\sqrt{n}\right)^k = x_k - y_k\sqrt{n}.$$

To see why this is so, imagine the binomial expansions:

$$\left(x_1 + y_1\sqrt{n}\right)^k = x_1^k + {}^kC_1 x_1^{k-1} y_1\sqrt{n} + {}^kC_2 x_1^{k-2}\left(y_1\sqrt{n}\right)^2 + {}^kC_3 x_1^{k-3}\left(y_1\sqrt{n}\right)^3 + \cdots$$

and

$$\left(x_1 - y_1\sqrt{n}\right)^k = x_1^k - {}^kC_1 x_1^{k-1} y_1\sqrt{n} + {}^kC_2 x_1^{k-2}\left(y_1\sqrt{n}\right)^2 - {}^kC_3 x_1^{k-3}\left(y_1\sqrt{n}\right)^3 + \cdots.$$

We wish to collate the terms which do not involve $\sqrt{n}$ and those which do. The terms which do not involve $\sqrt{n}$ are those where $\sqrt{n}$ is raised to an even power. These are identical in the two expansions, and so if they sum to x_k in one they do likewise in the other. The terms in which $\sqrt{n}$ is raised to an odd power are the same in the two expansions except, in the second, every sign has been changed. So in the first expansion they sum to $y_k\sqrt{n}$ and in the second to $-y_k\sqrt{n}$.

Therefore,

$$\begin{aligned} x_k^2 - ny_k^2 &= \left(x_k + y_k\sqrt{n}\right)\left(x_k - y_k\sqrt{n}\right) \\ &= \left(x_1 + y_1\sqrt{n}\right)^k \left(x_1 - y_1\sqrt{n}\right)^k \\ &= (x_1^2 - ny_1^2)^k = 1^k = 1. \end{aligned}$$

It remains to show that the solutions $x = x_k$, $y = y_k$ obtained in this way are the only solutions. In the hope of reaching a contradiction suppose that $x = a$, $y = b$ is some other positive solution. Then, since $x = x_1$, $y = y_1$ is the smallest positive solution and $x_1 + y_1\sqrt{n} > 1$, there exists a positive integer k such that

$$\left(x_1 + y_1\sqrt{n}\right)^k < a + b\sqrt{n} < \left(x_1 + y_1\sqrt{n}\right)^{k+1}.$$

We now wish to multiply through this inequality by $\left(x_1 - y_1\sqrt{n}\right)^k$. Notice that

$$x_1 - y_1\sqrt{n} = \frac{1}{x_1 + y_1\sqrt{n}} > 0,$$

and so multiplication by the positive amount $\left(x_1 - y_1\sqrt{n}\right)^k$ will preserve order. Moreover, we have seen that this multiplier is equal to $x_k - y_k\sqrt{n}$, and so

$$\left(x_1 + y_1\sqrt{n}\right)^k \left(x_1 - y_1\sqrt{n}\right)^k < \left(a + b\sqrt{n}\right)\left(x_k - y_k\sqrt{n}\right) < \left(x_1 + y_1\sqrt{n}\right)^{k+1}\left(x_1 - y_1\sqrt{n}\right)^k.$$

This expression simplifies to

$$1 < \left(a + b\sqrt{n}\right)\left(x_k - y_k\sqrt{n}\right) < x_1 + y_1\sqrt{n}.$$

Now $\left(a + b\sqrt{n}\right)\left(x_k - y_k\sqrt{n}\right) = c + d\sqrt{n}$, where $c = ax_k - nby_k$ and $d = bx_k - ay_k$. A little algebraic manipulation confirms that

$$c^2 - nd^2 = (x_k^2 - ny_k^2)(a^2 - nb^2) = 1.$$

Therefore $x = c$, $y = d$ is a solution of the equation and the established inequality, which can now be written as

$$1 < c + d\sqrt{n} < x_1 + y_1\sqrt{n},$$

will contradict the fact that $x_1 + y_1\sqrt{n}$ is the fundamental solution, provided we can confirm that c and d are both positive. For this result, note that since $\left(c + d\sqrt{n}\right)\left(c - d\sqrt{n}\right) = 1$ and $c + d\sqrt{n} > 1$, we have $0 < c - d\sqrt{n} < 1$. Therefore

$$2c = \left(c + d\sqrt{n}\right) + \left(c - d\sqrt{n}\right) > 1 + 0, \quad \text{giving } c > 0,$$

and

$$2d\sqrt{n} = \left(c + d\sqrt{n}\right) - \left(c - d\sqrt{n}\right) > 1 - 1, \quad \text{giving } d > 0.$$

So the proof is complete. ■

Although we have concentrated on the equation $x^2 - ny^2 = 1$ we could apply our line of proof in Theorem 1.3 to the allied equation $x^2 - ny^2 = -1$. This equation will not have any solutions unless the cycle in the ICF of $\sqrt{n}$ has odd length. For those n which do admit solutions we have the following result. Its proof, which we shall not spell out here, involves no more than retracing the steps of the above proof and changing signs where appropriate.

Theorem 1.4

Suppose that the ICF of $\sqrt{n}$ has a cycle of odd length s. Let $x_1 = p_s$, $y_1 = q_s$, where the convergent $C_s = \dfrac{p_s}{q_s}$, and let x_k and y_k be given by

$$x_k + y_k\sqrt{n} = \left(x_1 + y_1\sqrt{n}\right)^k.$$

Then, for all integers $k \geq 1$, $x = x_{(2k-1)s}$, $y = y_{(2k-1)s}$ is a solution of $x^2 - ny^2 = -1$ and $x = x_{2ks}$, $y = y_{2ks}$ is a solution of $x^2 - ny^2 = 1$. Moreover, all solutions of $x^2 - ny^2 = \pm 1$ are given in this way.

Problem 1.4

Given that $\sqrt{17} = [4, \langle 8 \rangle]$ find the smallest solution of $x^2 - 17y^2 = -1$ and hence find two positive solutions of $x^2 - 17y^2 = 1$.

Problem 1.5

For a number to be both triangular and square there must exist integers m and n such that $\dfrac{n(n+1)}{2} = m^2$. Substituting $n = \dfrac{x-1}{2}$ and $m = \dfrac{y}{2}$ this equation becomes $x^2 - 2y^2 = 1$. Use the solutions of this Pell's equation to find five triangular squares.

Note that $\sqrt{2} = [1, \langle 2 \rangle]$.

2 THE PYTHAGOREAN EQUATION

2.1 Primitive Pythagorean triples

The theorem of Pythagoras, which states that the sum of the squares on the two shorter sides of a right-angled triangle is equal to the square on the hypoteneuse, is arguably the most celebrated theorem in mathematics. But in addition to being a classical theorem of geometry, it presents challenges for the number theorist when the underlying equation $x^2 + y^2 = z^2$ is considered as a Diophantine equation. We refer to this as the *Pythagorean equation*. Interest focuses on right-angled triangles all three of whose sides are positive integers. The Pythagorean equation $x^2 + y^2 = z^2$ certainly has solutions in positive integers, including the well-known one which has the smallest value of z, namely $3^2 + 4^2 = 5^2$. It is also immediate from this one solution that the equation has infinitely many solutions since, for any positive integer k, $(3k)^2 + (4k)^2 = (5k)^2$. But there is a sense in which this infinite family of solutions is really just one solution. Geometrically, the corresponding triangles are similar; we have the one basic triangle with sides 3, 4 and 5 and the others are obtained by scaling each side by factor k.

We shall continue the geometrical analogy by referring to the value of z in any solution as being the *hypoteneuse.*

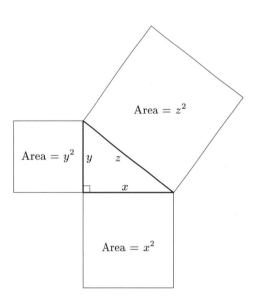

Figure 2.1   Pythagoras' Theorem: $x^2 + y^2 = z^2$

Before progressing further let us introduce some terminology.

Definition 2.1 Pythagorean triples

A *Pythagorean triple* is a triple (x, y, z) of positive integers such that $x^2 + y^2 = z^2$. The triple (x, y, z) is said to be *primitive* if $\gcd(x, y) = 1$.

A triple is ordered so that the hypoteneuse is always the third member.

Notice that the condition $\gcd(x, y) = 1$ guarantees that x and y have no common divisor greater than 1, and consequently the primitive triple (x, y, z) cannot be a scaling of some smaller triple. Moreover, the condition $\gcd(x, y) = 1$ carries the implications that $\gcd(x, z) = \gcd(y, z) = 1$, since from the equation $x^2 + y^2 = z^2$ it is readily shown that any common divisor of two of the variables must divide the third.

Problem 2.1 _____

Which, if any, of the following are Pythagorean triples? For each
Pythagorean triple, decide whether or not it is primitive.

(a) $(10, 8, 6)$ (b) $(12, 9, 15)$ (c) $(6, 7, 8)$ (d) $(5, 12, 13)$

(e) $(24, 33, 41)$

_____

We now have two primitive Pythagorean triples, namely $(3, 4, 5)$ and
$(5, 12, 13)$. Are there any more? Table 2.1 displays the start of a curious
infinite family of primitive solutions.

Table 2.1   A family of primitive Pythagorean triples

x	y	z
21	220	221
201	20 200	20 201
2 001	2 002 000	2 002 001
20 001	200 020 000	200 020 001
200 001	20 000 200 000	20 000 200 001
2 000 001	2 000 002 000 000	2 000 002 000 001
20 000 001	200 000 020 000 000	200 000 020 000 001
. . .	. . .	. . .

The Babylonians, some 3500 years ago, were aware of many primitive
solutions and Pythagoras himself is credited with the first infinite family of
solutions given by

$$x = 2k + 1, \quad y = 2k^2 + 2k, \quad z = 2k^2 + 2k + 1, \quad \text{for any integer } k \geq 1.$$

(The solutions displayed in Table 2.1 are taken from this family by
choosing k to be 10, 100, 1000, ...) But this infinite family does not
exhaust all primitive solutions; for example it does not include the solution
$(8, 15, 17)$. The first complete solution of the Pythagorean equation
appeared in Euclid's *Elements*. We aim to reproduce that solution here, but
first there are a few points to be clarified.

If (x, y, z) is any Pythagorean triple then x and y cannot both be odd for if
they were then

$$z^2 = x^2 + y^2 \equiv 1 + 1 \equiv 2 \pmod 4,$$

which is impossible as all squares are congruent modulo 4 to either 0 or 1. It
follows that in a primitive triple x and y have opposite parity because the
further condition that $\gcd(x, y) = 1$ ensures that they cannot both be even.

If integers x and y are both odd or
both even they are said to have the
same parity; otherwise they are
said to have *opposite parity*.

If (x, y, z) is a primitive Pythagorean triple then so too is (y, x, z) because
the x and y values can certainly be interchanged. But, to all intents and
purposes, the triples (x, y, z) and (y, x, z) lead to what is really the same
solution of the Pythagorean equation. To overcome the need to distinguish
between these two equivalent solutions we shall choose to write the even
member first in any primitive triple. We adopt the following convention.

> **Convention for Pythagorean triples**
>
> If (x, y, z) is any primitive Pythagorean triple, x is even while y and z
> are both odd.

With the preparation now complete we are ready for the main theorem of this section.

> ### Theorem 2.1 *Primitive solutions of the Pythagorean equation*
>
> The primitive Pythagorean triples are the triples
>
> $$(2mn,\ m^2 - n^2,\ m^2 + n^2),$$
>
> where m and n are relatively prime positive integers of opposite parity and with $m > n$.

Observe that the even value, $x = 2mn$, is always a multiple of 4.

Before embarking on the proof let us clarify the claim of the theorem by looking at an example of its use.

Example 2.1

Find all primitive Pythagorean triples $(60, y, z)$.

We seek relatively prime integers m and n such that $2mn = 60$, m and n have opposite parity and $m > n$. There are four pairs of positive integers satisfying $mn = 30$ with $m > n$ and in all four cases m and n are relatively prime and have opposite parity:

$$m = 30,\ n = 1 \text{ giving } (60, 899, 901);$$
$$m = 15,\ n = 2 \text{ giving } (60, 221, 229);$$
$$m = 10,\ n = 3 \text{ giving } (60, 91, 109);$$
$$m = 6,\ n = 5 \text{ giving } (60, 11, 61).$$

These four are the only primitive Pythagorean triples $(60, y, z)$. ♦

Proof of Theorem 2.1

We must first show that the given triple is a primitive Pythagorean triple. Substituting $x = 2mn$, $y = m^2 - n^2$ and $z = m^2 + n^2$ in the Pythagorean equation we have

$$x^2 + y^2 = (2mn)^2 + (m^2 - n^2)^2$$
$$= m^4 + 2m^2n^2 + n^4 = (m^2 + n^2)^2 = z^2.$$

So this is indeed a Pythagorean triple. To establish the primitive property suppose to the contrary that p is a prime divisor of both $m^2 - n^2$ and $m^2 + n^2$. Then p divides both the sum $(m^2 + n^2) + (m^2 - n^2) = 2m^2$ and the difference $(m^2 + n^2) - (m^2 - n^2) = 2n^2$. But $p \neq 2$ (since y is odd) and we conclude that p divides m and p divides n. As m and n are given to be relatively prime we have the required contradiction, and so the triple is primitive.

Conversely we have to show that every primitive Pythagorean triple is of the stated form. With this goal in mind suppose that (x, y, z) is *any* primitive Pythagorean triple. As y and z are both odd, we can define integers

$$s = \frac{z + y}{2} \text{ and } t = \frac{z - y}{2}.$$

We note that s and t are relatively prime because any common divisor would also divide $s + t = z$ and $s - t = y$.

Then from the Pythagorean equation

$$x^2 = z^2 - y^2 = (z + y)(z - y) = 2s \times 2t$$

we obtain

$$\left(\frac{x}{2}\right)^2 = st.$$

As x is known to be even this equation confirms that the product st is a perfect square. Therefore, since s and t are relatively prime, they must each be squares. So if we now write $s = m^2$ and $t = n^2$ and substitute back:

$$x^2 = 4st = 4m^2n^2, \quad \text{giving } x = 2mn;$$
$$y = s - t = m^2 - n^2;$$
$$z = s + t = m^2 + n^2.$$

This follows from the work of *Unit 2*, but see also Problem 2.4 at the end of this subsection.

Finally, note that $\gcd(m, n) = 1$, because any common divisor of m and n would necessarily be a common divisor of the relatively prime integers $s = m^2$ and $t = n^2$. Moreover m and n must have opposite parity for otherwise y and z would be even.

The triple (x, y, z) has thus been expressed in the required form. ∎

All Pythagorean triples can be obtained by scaling primitive ones; that is, any Pythagorean triple is of the form (kx, ky, kz), where (x, y, z) is a primitive triple. Notice that our convention for primitive triples carries over to all triples. For example, consider the triples $(6, 8, 10)$ and $(8, 6, 10)$ which we wish to regard as being the same. Since the underlying primitive triple is $(4, 3, 5)$ we shall conventionally write this multiple as $(8, 6, 10)$, rather than $(6, 8, 10)$. In other words, the first member in the triple is the multiple of the even member in the underlying primitive triple. With this clarified we can now record the following Corollary.

Corollary **All solutions of the Pythagorean equation**

The Pythagorean triples (x, y, z) are given by

$$x = 2kmn, \quad y = k(m^2 - n^2), \quad z = k(m^2 + n^2),$$

where $k \geq 1$ is any integer and m and n are relatively prime positive integers with opposite parity and $m > n$.

This implies that there are no Pythagorean triples (x, y, z) with $x = y$.

In the next example we make use of Theorem 2.1 to begin an enumeration of all the primitive Pythagorean triples.

Example 2.2

There are 16 primitive Pythagorean triples (x, y, z) with hypotenuse $z < 100$. Find them.

We find primitive triples by listing pairs of relatively prime integers m and n which have opposite parity and with $m > n$. For each such pair the corresponding triple is calculated from the formulae in Theorem 2.1. For example, taking $m = 7$ each of the values $n = 2$, 4 and 6 will lead to a primitive triple.

For $m = 7$ and $n = 1$, 3 or 5 the resulting triple will be Pythagorean but will not be primitive because m and n have the same parity.

The condition that $z < 100$ amounts to $m^2 + n^2 < 100$. Hence $m \leq 9$ and, for example, the triple resulting from $m = 8$, $n = 7$ is not wanted because its hypotenuse exceeds 100. The 16 solutions are as follows.

Table 2.2 Primitive triples with sides not exceeding 100

m	n	triple	m	n	triple
2	1	$(4, 3, 5)$	7	2	$(28, 45, 53)$
3	2	$(12, 5, 13)$	7	4	$(56, 33, 65)$
4	1	$(8, 15, 17)$	7	6	$(84, 13, 85)$
4	3	$(24, 7, 25)$	8	1	$(16, 63, 65)$
5	2	$(20, 21, 29)$	8	3	$(48, 55, 73)$
5	4	$(40, 9, 41)$	8	5	$(80, 39, 89)$
6	1	$(12, 35, 37)$	9	2	$(36, 77, 85)$
6	5	$(60, 11, 61)$	9	4	$(72, 65, 97)$

◆

Problem 2.2 ───────────────────────────────────

Find all primitive Pythagorean triples $(72, y, z)$.

Problem 2.3 ───────────────────────────────────

Find all Pythagorean triples, primitive or not, in which one of the sides is 30.

Problem 2.4 ───────────────────────────────────

Give a proof of the step assumed in Theorem 2.1 that, if a product of two relatively prime integers is a square, then each of the integers must be square.

───

2.2 Special Pythagorean triples

From the classification of Theorem 2.1 it is not difficult to see exactly which integers can arise as members of a primitive Pythagorean triple. If we choose m and n to be consecutive integers $m = k + 1$, $n = k$ (for any $k \geq 1$) then they are relatively prime and of opposite parity and so give the primitive triple $(2k^2 + 2k,\ 2k + 1,\ 2k^2 + 2k + 1)$. The middle value, $2k + 1$, shows that every odd integer from 3 onwards can be a member of a primitive Pythagorean triple. As far as even integers are concerned, the even member is $x = 2mn$, where m and n have opposite parity, and consequently x is a multiple of 4. This means that no integer which is congruent modulo 4 to 2 can be a member of a primitive Pythagorean triple. However the choice $m = 2k$ and $n = 1$ gives the triple $(4k,\ 4k^2 - 1,\ 4k^2 + 1)$ and proves that every multiple of 4 occurs as a member of a primitive Pythagorean triple. So every integer $n > 1$, with the exception of those $n \equiv 2 \pmod 4$, occurs in some primitive Pythagorean triple.

This is the family discovered by Pythagoras which we mentioned earlier, but with x and y values interchanged to fit our convention.

What about occurrences of primes? Not all three members of a Pythagorean triple can be prime because one member is a multiple of 4. On the other hand, Table 2.2 reveals several instances where both the odd members of a primitive triple are prime. Suppose that the y-value in a triple is prime, say $y = p$. Then, from $x^2 + p^2 = z^2$ we have

$$p^2 = z^2 - x^2 = (z + x)(z - x).$$

As $z > x > 0$, the two terms on the right of this equation are distinct and positive. Now there is only one way that p^2 can be written as the product of unequal positive divisors, namely $p^2 = p^2 \times 1$. Hence

$$z + x = p^2 \quad \text{and} \quad z - x = 1,$$

and solving these equations for x and z gives the primitive triple

$$\left(\frac{p^2 - 1}{2},\ p,\ \frac{p^2 + 1}{2} \right).$$

As $p^2 \equiv 1 \pmod 8$ we have confirmation that $\dfrac{p^2 - 1}{2}$ is divisible by 4.

So a Pythagorean triple contains two prime members whenever the prime p is such that $\dfrac{p^2 + 1}{2}$ is also prime. Many examples of such pairs of primes are known, but it is one more of the many unsolved problems of number theory as to whether or not there are infinitely many such related pairs of primes.

To round off this section let us look at one more associated problem. The triple $(4, 3, 5)$ has the property that its three sides are consecutive integers. It is not difficult to show that no other triple can have this property. There are certainly instances where two of the members are consecutive. As we discovered above, each triple in the infinite family of primitive triples of the form $(2k^2 + 2k,\ 2k + 1,\ 2k^2 + 2k + 1)$ has the even member and the hypoteneuse being consecutive integers. The remaining question concerns the possibility of the two shorter sides being consecutive. Table 2.2 uncovered two triples with this propety, namely $(4, 3, 5)$ and $(20, 21, 29)$. Are there any more?

Example 2.3

Find primitive triples (x, y, z) in which x and y are consecutive integers.

Adopting the notation of Theorem 2.1, suppose that $x = 2mn$ and $y = m^2 - n^2$ are consecutive integers. That is,

$$m^2 - n^2 = 2mn \pm 1.$$

Taking all terms involving m to the left-hand side and completing the square gives

$$(m - n)^2 = 2n^2 \pm 1.$$

Now the substitution $r = m - n$ yields

$$r^2 - 2n^2 = \pm 1.$$

This should look familiar from the work of the previous section. All solutions in r and n arise from convergents in the ICF of $\sqrt{2} = [1, \langle 2 \rangle]$ which begin

$$\frac{1}{1}, \frac{3}{2}, \frac{7}{5}, \frac{17}{12}, \frac{41}{29}, \ldots$$

As the cycle of $\sqrt{2}$ has length 1, each convergent $\dfrac{p_k}{q_k}$ gives a solution $r = p_k$, $n = q_k$ of one of the equations $r^2 - 2n^2 = \pm 1$, and the required value of m is recovered by means of $m = r + n$. The first five convergents lead to triples as follows.

Theorem 1.2

r	n	m	x	y	z
1	1	2	4	3	5
3	2	5	20	21	29
7	5	12	120	119	169
17	12	29	696	697	985
41	29	70	4060	4059	5741

♦

Diversion

As a diversion we extend the table begun in Example 2.3 by listing the first twenty Pythagorean triples with consecutive shorter sides.

4	3	5
20	21	29
120	119	169
696	697	985
4060	4059	5741
23660	23661	33461
137904	137903	195025
803760	803761	1136689
4684660	4684659	6625109
27304196	27304197	38613965
159140520	159140519	225058681
927538920	927538921	1311738121
5406093004	5406093003	7645370045
31509019100	31509019101	44560482149
183648021600	183648021599	259717522849
1070379110496	1070379110497	1513744654945
6238626641380	6238626641379	8822750406821
36361380737780	36361380737781	51422757785981
211929657785304	211929657785303	299713796309065
1235216565974040	1235216565974041	1746860020068409

Problem 2.5 _____

Find all Pythagorean triples in which the three sides are in arithmetic progression.

Problem 2.6 _____

Show that there are infinitely many Pythagorean triples in which the shorter sides are consecutive triangular numbers T_k and T_{k+1}. *Hint*: If $(m, m+1, n)$ is a Pythagorean triple try the triangular numbers T_{2m} and T_{2m+1}.

Diversion

It is possible to have primitive Pythagorean triples all three of whose sides are triangular numbers. For example,

$$T_{132}^2 + T_{143}^2 = T_{164}^2.$$

It is not known whether infinitely many such triples exist.

Is it possible to have a primitive Pythagorean triple all three of whose members are squares?

---

3 FERMAT'S LAST THEOREM

3.1 Some history of the Last Theorem

Fermat's refusal to publish his discoveries led to one of the great stories in the history of mathematics. As well as writing letters to his friends about various results, he formed the habit of jotting notes in the margins of his reference books. These notes to himself comprised brief summaries of his discoveries. Some five years after his death his copy of Bachet's *Diophantus* was found to contain mention of many results from number theory. On the page dealing with the Pythagorean equation Fermat had added in the margin:

> *On the other hand it is impossible to write a cube as the sum of two cubes, a fourth power as the sum of two fourth powers and, in general, any power higher than the second as the sum of two similar powers. I have discovered a truly marvellous proof of this, but the margin is too small to contain it.*

In this, Fermat was claiming to have proved that the Diophantine equation $x^n + y^n = z^n$ has no positive solutions for $n \geq 3$. No evidence of how Fermat reached his conclusions survives, and despite three and a half centuries of continual efforts to solve the problem, nobody managed to prove, or disprove, Fermat's assertion. The difficulties encountered in attempting a general proof have convinced many mathematicians and historians that Fermat was mistaken and did not really have a proof. Measured against this is the fact that Fermat was a phenomenal mathematician and nothing which he ever claimed to have proved has subsequently been disproved. In the single instance in which he erroneously claimed something to be true, namely that the Fermat numbers are all prime, he left his contemporaries in no doubt that he was unable to prove it.

As this was the last of Fermat's results awaiting proof or refutation it has become known colloquially as *Fermat's Last Theorem*. Strictly speaking, it should not be a theorem until a proof has been found, and some texts refer to it as Fermat's Conjecture. However, for good reasons which we shall give shortly, we shall use the former, more notorious name.

Consider the equation $x^n + y^n = z^n$, for $n \geq 3$, and suppose the exponent n is composite, say $n = kr$. The equation can then be rewritten as

$$(x^k)^r + (y^k)^r = (z^k)^r.$$

It follows that if $x = x_0$, $y = y_0$, $z = z_0$ is a solution of $x^n + y^n = z^n$ then

$$x = x_0^k, \quad y = y_0^k, \quad z = z_0^k$$

is a solution of $x^r + y^r = z^r$. Conversely, if it can be established that $x^r + y^r = z^r$ has no positive solution then it follows that $x^n + y^n = z^n$ must have no positive solutions.

Now every integer $n \geq 3$ is divisible by some odd prime or by 4. The proof of Fermat's Last Theorem can therefore be broken down to two tasks. If each of the following can be proved

- $x^4 + y^4 = z^4$ has no positive solution
- $x^p + y^p = z^p$ has no positive solution for any odd prime p

then it will follow that $x^n + y^n = z^n$ has no positive solution for any $n \geq 3$.

Fermat himself did leave us an elegant proof for the case of the exponent 4, which we shall present in this section. The real challenge, however, is in proving Fermat's Last Theorem for odd prime exponents. In 1770 Euler resolved the case $p = 3$, although his proof was not quite complete. Half a century later Dirichlet and Legendre gave independent proofs for the case $p = 5$ and in 1835 Lamé supplied a proof for the case $p = 7$. But the proofs of these cases adopted very different approaches and it was becoming clear that to reach a proof applicable for all odd primes p some new approach was needed. Lamé thought he had made the breakthrough in 1847. In attempting to solve $x^p + y^p = z^p$, instead of working within the set of integers Lamé extended his system by looking at numbers of the form $m + n\alpha^r$, where m and n are integers, $\alpha^p = 1$ and $0 \leq r < p$. He thought he had succeeded in proving that the equation had no non-trivial solution in this number system, and therefore no positive integer solution. Alas, he made one serious error. He assumed that numbers in his new system would factorize uniquely. They do not, for the equivalent of the Fundamental Theorem of Arithmetic, which leads to unique factorization in the integers, does not hold in this ring of numbers.

These numbers are called *algebraic numbers* and Lamé had invented the first example of what algebraists today call a *ring*.

The German mathematician Kummer, who may have been simultaneously pursuing similar lines to Lamé, did achieve a significant advance. To overcome the lack of unique factorization his remedy was to extend his number system in a different way inventing the so-called *ideal prime divisors*. Kummer successfully proved Fermat's Last Theorem for the case of exponents which he classed as *regular primes*. This was no mean achievement for the regular primes include all primes less than 100 with the three exceptions of 37, 59 and 67. Kummer believed that he had proved Fermat's Last Theorem for infinitely many primes but it remains another unsolved problem of mathematics as to whether or not there are infinitely many of Kummer's regular primes; ironically, however, it has been known since 1915 that there are infinitely many irregular ones.

By developing Kummer's ideas, mathematicians have settled more and more cases of Fermat's Last Theorem over the years. The arrival of electronic calculators inevitably speeded up discoveries and, for example, by 1980 Fermat's Last Theorem had been shown to hold true for all odd prime exponents not exceeding 125 000. By 1994 that had been advanced to all prime exponents not exceeding 4 000 000.

One significant breakthrough was achieved in 1983 by a young German mathematician, Gerd Faltings. He proved that for any given n there is at most a finite number of solutions of $x^n + y^n = z^n$. All in all, the overwhelming numerical evidence left little doubt that the result asserted in Fermat's Last Theorem must be true, but that elusive complete proof was still missing.

Over the years a number of prizes have been offered for the first general proof of Fermat's Last Theorem. The Academy of Science at Paris offered prizes in 1823 and again in 1850, and the Academy of Brussels put up a prize in 1883. But most memorable, in 1908 an enormous prize of 100 000 marks was bequeathed to the Academy of Science at Gottingen for the first complete solution of Fermat's Last Theorem. The entry condition that the solution must be printed no doubt deterred many but, nevertheless, over 1000 solutions were submitted during the next four years. Inevitably these submissions were either erroneous or incomplete. Landau, a distinguished German number theorist of that period, was burdened with the task of checking the submissions. It is reported that he had postcards printed which read "Dear Sir or Madam, Your attempted proof of Fermat's Last Theorem has been received and is returned herewith. The first mistake is on page ... , line ... ." The job of filling in the missing entries was given to Landau's research students!

German inflation of the 1920's reduced the value of the Gottingen prize to virtually nothing, but the pursuit of a proof of Fermat's Last Theorem has not waned. In 1988 the mathematical world was excited by the news that a Japanese mathematician, Yoichi Miyaoka, had resolved Fermat's Last Theorem. Alas, as so often before, holes were found in the purported proof. Then in 1994 the distinguished British mathematician Andrew Wiles, now working at Princeton University, gave a series of lectures at Cambridge University which appeared to have culminated in a proof of Fermat's Last Theorem. Although holes were initially found in this proof, it was believed that they could be plugged. The proof is of some 1000 pages and takes an incredibily circuitous route to the result, straying a long way outside Elementary Number Theory. But, at the time of writing this course, the mathematical world is holding its breath in the growing belief that Wiles has, at long last, put proof of Fermat's Last Theorem to rest.

3.2 The equation $x^4 + y^4 = z^4$

In the remainder of this section we shall confine our attention to the special case of Fermat's Last Theorem for exponent 4 and some related Diophantine equations. We shall use Fermat's method and, just as he did, we shall prove the stronger result that $x^4 + y^4 = z^2$ has no positive solution. The proof uses our solution to the Pythagorean equation together with a powerful technique known as Fermat's *method of infinite descent*.

The idea behind the method of infinite descent as applied to this problem is as follows. We establish that from any positive solution $x = x_k$, $y = y_k$, $z = z_k$ of the Diophantine equation $x^4 + y^4 = z^2$ we can always find another 'smaller' positive solution $x = x_{k+1}$, $y = y_{k+1}$, $z = z_{k+1}$, where by smaller we mean that $0 < z_{k+1} < z_k$. This process of being able to go from one positive solution to a smaller one is the key stage of the method, and is called the *descent step*.

Now suppose that the equation does have a positive solution $x = x_1$, $y = y_1$, $z = z_1$. The established descent step would then tell us that, from this, we have a smaller positive solution, and from this second solution a still smaller third positive solution, from which we get a smaller fourth positive solution, and so on. Focussing on the z-values we have an unending decreasing sequence of positive integers

$$z_1 > z_2 > z_3 > z_4 > \cdots > 0.$$

As there are only finitely many positive integers which are less than z_1 this is plainly impossible. Hence the one assumption that we have made, namely that one positive solution exists, is contradicted.

We have chosen to focus on the variable z here. But notice that descent could be achieved on any of the variables. For example, if we established that any positive solution of the equation must give rise to another positive solution with smaller value for y, then the conclusion that no solution can exist follows in the same way.

What is really being brought into play here is the Well-Ordering Principle which asserts that any non-empty set of positive integers must have a least member. The descent step appears to construct a set of positive integers which does not have a least member, and so it has to be the empty set.

> ### Theorem 3.1
>
> The Diophantine equation $x^4 + y^4 = z^2$ has no positive solutions.

The proof of Theorem 3.1, and that of Theorem 3.2 which follows shortly, involve some tricky algebra. You are not expected to master the details of these proofs, but you should read each of them carefully to see how the method of infinite descent is used.

Proof of Theorem 3.1

Suppose to the contrary that $x^4 + y^4 = z^2$ has a positive solution $x = x_1$, $y = y_1$, $z = z_1$. If $\gcd(x_1, y_1) = d > 1$ then putting $x_1 = dx'$ and $y_1 = dy'$ gives $z_1^2 = d^4(x'^4 + y'^4)$, from which it follows that d^2 divides z_1 so that $z_1 = d^2 z'$, for some integer z'. But then $x'^4 + y'^4 = z'^2$, where $\gcd(x', y') = 1$ and $z' < z$. This argument shows that we may assume that $\gcd(x_1, y_1) = 1$ for otherwise any common divisor can first be cancelled leaving another, smaller, solution to the same equation.

Writing the equation as $(x_1^2)^2 + (y_1^2)^2 = z_1^2$ we observe that (x_1^2, y_1^2, z_1) is a primitive Pythagorean triple and so, by Theorem 2.1,

$$x_1^2 = 2mn, \quad y_1^2 = m^2 - n^2, \quad z_1 = m^2 + n^2,$$

where m and n are relatively prime positive integers of opposite parity with $m > n$. In fact n must be even and m odd, for otherwise we have

$$y_1^2 = m^2 - n^2 \equiv 0 - 1 \equiv 3 \pmod{4},$$

which is impossible because any square is congruent modulo 4 to either 0 or 1.

Putting $n = 2r$ the equation for x_1 becomes

$$\left(\frac{x_1}{2}\right)^2 = mr,$$

where m and $r = \dfrac{n}{2}$ are relatively prime. Hence m and r are each squares, say $m = s^2$ and $r = t^2$.

Returning to the equation $y_1^2 = m^2 - n^2$, we see that (n, y_1, m) is a primitive Pythagorean triple and so

$$n = 2uv, \quad y_1 = u^2 - v^2, \quad m = u^2 + v^2,$$

where u and v are relatively prime positive integers of opposite parity with $u > v$.

Now $n = 2r = 2t^2$, and so the first of these equations becomes $uv = t^2$. It follows that that u and v are each squares, say $u = x_2^2$ and $v = y_2^2$. Feeding these facts into the equation $m = u^2 + v^2$ gives

$$x_2^4 + y_2^4 = m = s^2 (= z_2^2, \text{ say}),$$

Infinite descent bears some resemblance to mathematical induction. The descent step, like the induction step, sets up an unending chain of implications of the form 'if one is true then so too is the next'. The difference is that this time we know that no such infinite chain of true statements can exist and we are forced to conclude that the chain cannot be initiated; it has no basis.

Rather than use the general suffix k as in the preamble, we establish the descent step here by showing that any positive solution x_1, y_1, z_1 must lead to a positive solution x_2, y_2, z_2 in which $0 < z_2 < z_1$.

If needed we may interchange x_1 and y_1 to ensure that x_1 is even.

revealing another solution of the original equation. But this solution is 'smaller' since

$$0 < z_2 = s \le m < m^2 + n^2 = z_1.$$

This completes the descent step; any positive solution gives rise to a smaller, positive one. As this is impossible, the assumption that there exists a positive solution is contradicted. ∎

As any fourth power is necessarily a square, any positive solution of $x^4 + y^4 = z^4$ would contradict the result of Theorem 3.1. Hence

---

Corollary to Theorem 3.1

The Diophantine equation $x^4 + y^4 = z^4$ has no positive solutions.

The method of infinite descent used in the proof of Theorem 3.1 can be applied to many Diophantine equations, almost invariably with the purpose of showing that the equation has no positive solutions. Here is a simpler example.

Example 3.1

Show that the Diophantine equation $x^3 + 3y^3 = 9z^3$ has no positive solutions.

Aiming for a contradiction suppose that $x_1^3 + 3y_1^3 = 9z_1^3$, where x_1, y_1 and z_1 are positive integers. As the prime 3 divides the right-hand side and one of the terms on the left of this equation, it must divide the remaining term. That is, 3 divides x_1^3 and so 3 divides x_1. Substituting $x_1 = 3x_2$ in the equation gives

$$27x_2^3 + 3y_1^3 = 9z_1^3$$

and, dividing throughout by 3,

$$9x_2^3 + y_1^3 = 3z_1^3.$$

Repetition of the same reasoning now shows that 3 divides y_1 and putting $y_1 = 3y_2$ the equation becomes

$$9x_2^3 + 27y_2^3 = 3z_1^3; \quad \text{that is, } 3x_2^3 + 9y_2^3 = z_1^3.$$

This time we have 3 dividing z_1, and putting $z_1 = 3z_2$ the equation now becomes

$$3x_2^3 + 9y_2^3 = 27z_2^3; \quad \text{that is, } x_2^3 + 3y_2^3 = 9z_2^3.$$

At this point we have reached a second solution, $x = x_2$, $y = y_2$, $z = z_2$ of the original equation with $z_2 < z_1$. The descent step is therefore complete and the required contradiction is established. Thus $x^3 + 3y^3 = 9z^3$ has no positive solutions. ◆

Problem 3.1 ————————————————————————

Show that the Diophantine equation $x^4 + 2y^4 = 4z^4$ has no positive solutions.

————————————————————————

The following problem requires a slight variant on Fermat's method of infinite descent.

Problem 3.2

(a) Show that if $x^2 + y^2 + z^2 \equiv 0 \pmod 4$ then each of x, y and z is even.

(b) Show that in any positive solution $x = x_1$, $y = y_1$, $z = z_1$ of the Diophantine equation $x^2 + y^2 + z^2 = 2xyz$, each of x_1, y_1 and z_1 must be even. Putting $x_1 = 2x_2$, $y_1 = 2y_2$ and $z_1 = 2z_2$, show that x_2, y_2 and z_2 are also even positive integers.

(c) Prove that the Diophantine equation $x^2 + y^2 + z^2 = 2xyz$ has no positive solution by constructing from such a solution a strictly decreasing infinite sequence of even positive integers.

3.3 Related Diophantine equations

Fermat went on to apply his method of infinite descent to prove that the related equation $x^4 - y^4 = z^2$ has no positive solution. However we shall first establish Theorem 3.2 and then deduce Fermat's result as a corollary.

> ***Theorem 3.2***
>
> The Diophantine equation $x^4 + 4y^4 = z^2$ has no positive solution.

Proof of Theorem 3.2

Suppose that x_1, y_1 and z_1 are positive integers with $x_1^4 + 4y_1^4 = z_1^2$. Exactly as in the proof of Theorem 3.1 we may assume that $\gcd(x_1, y_1) = 1$, for otherwise cancellation of any common divisor leads to a smaller solution of the same equation.

With infinite descent in mind, we wish to show that the existence of this one positive solution necessarily gives rise to a one with a smaller z-value. We observe that x_1 and z_1 are of the same parity, and we first consider the case in which they are both even. Substituting $x_1 = 2x'$ and $z_1 = 2z'$ gives

$$(2x')^4 + 4y_1^4 = (2z')^2; \quad \text{that is,} \quad y_1^4 + 4x'^4 = z'^2$$

which is another positive solution of the equation with $z' < z$.

It remains to show that a solution in which x_1 and z_1 are both odd must likewise lead to another with smaller z-value. Writing the equation as $(x_1^2)^2 + (2y_1^2)^2 = z_1^2$ we recognize that $(2y_1^2, x_1^2, z_1)$ is a primitive Pythagorean triple and so, according to Theorem 2.1, there are relatively prime positive integers m and n of opposite parity and with $m > n$ such that

$$2y_1^2 = 2mn, \quad x_1^2 = m^2 - n^2, \quad z_1 = m^2 + n^2.$$

From $x_1^2 = m^2 - n^2$ we see that m is odd and n even, for the other way round would give $m^2 - n^2 \equiv 3 \pmod 4$ and this cannot be a square. And from $y_1^2 = mn$, the relatively prime integers m and n are each squares and so we can write $m = a^2$ and $n = (2b)^2$.

Looking again at $x_1^2 = m^2 - n^2$ with $\gcd(m, n) = 1$, we now see that (n, x_1, m) is another primitive Pythagorean triple, and so there exist relatively prime positive integers s and t, of opposite parity and with $s > t$ such that

$$n = 2st, \quad x_1 = s^2 - t^2, \quad m = s^2 + t^2.$$

Substituting $n = (2b)^2$ into the first of these equations gives $2b^2 = st$. Now one of s or t is even. If it is s, say $s = 2r$, then $b^2 = rt$ and as r and t are relatively prime each is a square; say $r = u^2$ (so that $s = 2u^2$) and $t = v^2$. The alternative that t is even leads, in exactly the same way, to $s = v^2$ and $t = 2u^2$. In either case, substituting into $m = s^2 + t^2$ gives

$$v^4 + 4u^4 = s^2 + t^2 = m = a^2.$$

The solution is $x = y_1$, $y = x'$, $z = z'$.

The triple is primitive because $\gcd(m, n) = 1$.

Thus we have another solution of the original equation. Moreover, putting $z_2 = a$ we have

$$0 < z_2 = a \leq m < m^2 + n^2 = z_1,$$

and so the z-value in this new solution is a smaller positive integer.

This completes the descent step and the proof. ∎

Corollary to Theorem 3.2

The Diophantine equation $x^4 - y^4 = z^2$ has no positive solution.

Proof of the Corollary

Suppose that x_1, y_1 and z_1 are positive integers with $x_1^4 - y_1^4 = z_1^2$. Squaring both sides of this equation and rearranging gives

$$z_1^4 + 4(x_1 y_1)^4 = (x_1^4 + y_1^4)^2.$$

Thus $x = z_1$, $y = x_1 y_1$, $z = x_1^4 + y_1^4$ is a positive solution of $x^4 + 4y^4 = z^2$. But this contradicts Theorem 3.2 and so the equation $x^4 - y^4 = z^2$ has no positive solution. ∎

We finish with a problem for you to attempt. The solution to this problem first appeared in the margin of Fermat's copy of *Diophantus*.

Problem 3.3

Let (x, y, z) be a Pythagorean triple. The area of the associated right-angled triangle is $\dfrac{xy}{2}$, and for this to be a square $xy = 2n^2$ for some integer n. Show that if there is a positive simultaneous solution to this equation and to $x^2 + y^2 = z^2$, then there is a positive solution of $a^4 - b^4 = c^2$, and hence deduce that no Pythagorean triangle can have an area which is a square.

---

4 SUMS OF SQUARES

4.1 Representing primes as sums of two squares

Another problem which attracted Fermat's attention concerned ways of expressing positive integers as sums of squares. For example, the integers from 1 to 9 can be written as sums of squares as follows.

$$1 = 1^2$$
$$2 = 1^2 + 1^2$$
$$3 = 1^2 + 1^2 + 1^2$$
$$4 = 2^2$$
$$5 = 2^2 + 1^2$$
$$6 = 2^2 + 1^2 + 1^2$$
$$7 = 2^2 + 1^2 + 1^2 + 1^2$$
$$8 = 2^2 + 2^2$$
$$9 = 3^2$$

The expressions given are certainly not unique; for example we could write

$$4 = 1^2 + 1^2 + 1^2 + 1^2 \quad \text{or} \quad 9 = 2^2 + 2^2 + 1^2.$$

But notice that the given expression for 7 as a sum of squares is the best we can do in the sense that 7 cannot be expressed as a sum of fewer than four squares. If you try continuing the list that we have started, with the aim of expressing each integer as a sum of as few squares as possible, you should find that each requires no more than four squares.

Amongst other questions that these observations might suggest are the following.

- Which integers can be expressed as a sum of two squares?
- Which integers can be expressed as a sum of two squares in a unique way?
- Which integers can be expressed as a sum of three squares?
- Can all positive integers be expressed as a sum of four squares?

We accept 0 as a square so that, for example, $4 = 2^2 + 0^2$ is a legitimate way of expressing 4 as a sum of two squares. However we ignore negative integers since $(-a)^2 = a^2$.

Fermat appears to have solved the first two of these problems. In letters to Mersenne he claimed to have a proof, using his descent method, of the key step that every prime of the form $4k + 1$ is expressible as a sum of two squares in a unique way. But yet again Fermat did not leave a copy of his proof and the mathematical world had to wait until 1747, when Euler provided one.

The first serious contribution to the third question belongs to Diophantus who conjectured that no number of the form $8k + 7$ can be expressed as a sum of three squares. Fermat appears to have been first to write down exact criteria for a number to be the sum of three squares, namely that the number must not be of the form $4^n(8k + 7)$ for non-negative integers k and n. Proof of this was provided by Legendre in 1798.

Having discovered which integers can, and which cannot, be expressed as a sum of two squares, and which can, and which cannot, be expressed as a sum of three squares, is it worth proceeding to solve the analogous problem for four, five, six, ... squares? Well yes it is, because at the next stage the sequence of investigations reaches a conclusion when we uncover the classic result that every positive integer is a sum of four squares. It is believed that, from the way he posed his questions in this area, the result was probably suspected by Diophantus, but it was first expressed formally by Bachet in 1621. Shortly after this, Fermat tackled the problem and (surprise, surprise) he claimed that he had a proof which used his descent method. Euler made various attempts at it over a period of more than 40 years but without success, which shows just how difficult it is. Eventually the four-square conjecture was proved in 1770 by Lagrange who acknowledged that ideas originating from Euler played a substantial part in his proof.

First we shall investigate the two-square problem. The following identity, which is easily established, is going to be crucial in what follows.

Important identity for two squares

$$(a^2 + b^2)(c^2 + d^2) = (ac + bd)^2 + (ad - bc)^2$$

This identity is of theoretical importance because it tells us that if two positive integers m and n can each be written as a sum of two squares then so too can their product mn. This means that to show that a given integer can be written as a sum of two squares it is sufficient to show that each prime in its decomposition can be expressed this way.

But the identity also has a practical use, as illustrated by the following example.

Example 4.1

Express 4420 as a sum of two squares.

$$4420 = 2^2 \times 5 \times 13 \times 17 = (2^2 + 0^2)(2^2 + 1^2)(3^2 + 2^2)(4^2 + 1^2)$$

Using the identity three times:

$$
\begin{aligned}
4420 &= (4^2 + 2^2)(3^2 + 2^2)(4^2 + 1^2) \\
&= (16^2 + 2^2)(4^2 + 1^2) \\
&= 66^2 + 8^2.
\end{aligned}
$$

This answer is not unique. It turns out that there are four ways of writing 4420 as a sum of two squares. The other ways can all be obtained from the identity applied to different terms. We could change the order of the bracketed terms, or switch the two terms within a bracket, or make use of negative values for a, b, c or d. Each of these variations is illustrated below where we discover the other three solutions.

$$
\begin{aligned}
4420 &= (4^2 + 2^2)(4^2 + 1^2)(2^2 + 3^2) \\
&= (4^2 + 2^2)(11^2 + 10^2) = 64^2 + 18^2 \\
4420 &= (4^2 + 2^2)(11^2 + (-10)^2) = 24^2 + (-62)^2 = 62^2 + 24^2 \\
4420 &= ((4^2 + 1^2)(2^2 + 1^2))((2^2 + 0^2)(2^2 + 3^2)) \\
&= (9^2 + 2^2)(4^2 + 6^2) = 48^2 + 46^2
\end{aligned}
$$
♦

Problem 4.1 _____

Find three ways of expressing 325 as a sum of two squares.

_____

Turning to the question of whether a positive integer n can be expressed as a sum of two squares, our important identity guides us to look at the primes occurring in its decomposition. Now primes can be classified into three types:

- the even prime $2 = 1^2 + 1^2$, which is a sum of two squares;

- odd primes of the form $4k + 1$;

- odd primes of the form $4k + 3$.

The third category presents little problem. As a square is congruent modulo 4 to either 0 or 1, a sum of two squares is congruent modulo 4 to one of 0, 1 or 2. So no prime of the form $4k + 3$ can be expressed as a sum of two squares.

That leaves just the middle category. We shall prove that all primes of the form $4k + 1$ can be written as a sum of two squares. It will then be a simple matter to complete the two-square problem. The result was first stated and proved by Fermat, and our proof follows his, using the descent method, but in a different way from our previous applications.

_____

Theorem 4.1 Primes expressed as a sum of two squares

A prime p can be expressed as a sum of two squares if, and only if, $p = 2$ or $p \equiv 1 \pmod 4$.

_____

All that remains to be done is to show that any prime $p \equiv 1 \pmod 4$ can be expressed as a sum of two squares. Before we present the proof formally let us explain how we are going to use the descent method. We start by assuming that some multiple of p can be expressed as a sum of two squares. More precisely, we assume that

$$mp = x^2 + y^2$$

has a solution for some $1 \le m < p$. If $m = 1$ we have the required expression for p as a sum of two squares. If $m > 1$, we go on to deduce from the above equation that a smaller multiple of p can be expressed as a sum of two squares, that is,

$$m_1 p = u^2 + v^2,$$

where $1 \leq m_1 < m$. This is our descent step. The difference this time is that we shall go on to show that the equation $mp = x^2 + y^2$ really does have a solution with $1 \leq m < p$. The descent step tells us that from this solution there is then a smaller solution, and from this a smaller one, and so on. The essence of the descent method is that such a process cannot go on forever, and so must terminate. The only way it can terminate is by descending to a solution with $m_1 = 1$ (so that the descent step cannot be applied again). Hence the existence of a solution to $mp = x^2 + y^2$ with $1 \leq m < p$ leads to the conclusion that $p = x^2 + y^2$ must have a solution.

Now for the details.

Proof of Theorem 4.1

Suppose that the equation $mp = x^2 + y^2$ has a solution in which $1 < m < p$. Let u and v be the least absolute residues modulo m of x and y respectively. That is,

$$u \equiv x \ (\text{mod } m), \quad v \equiv y \ (\text{mod } m), \quad -\frac{m}{2} < u, v \leq \frac{m}{2}.$$

Then

$$u^2 + v^2 \equiv x^2 + y^2 \equiv 0 \ (\text{mod } m),$$

and so

$$u^2 + v^2 = mr, \quad \text{for some integer } r \geq 0.$$

To establish the descent step we aim to show that rp is a sum of two squares with $1 \leq r < m$. First we check that r does lie in this range.

If $r = 0$ then $u = v = 0$ implying that m divides both x and y. But then, from $mp = x^2 + y^2$, we conclude that m divides p, which is plainly impossible. So

$$1 \leq r = \frac{u^2 + v^2}{m} \leq \frac{1}{m} \times \left(\frac{m^2}{4} + \frac{m^2}{4} \right) = \frac{m}{2} < m.$$

Now we must show that rp is a sum of two squares.

Multiplying together the equations $mp = x^2 + y^2$ and $mr = u^2 + v^2$, gives

$$m^2 rp = (x^2 + y^2)(u^2 + v^2) = (xu + yv)^2 + (xv - yu)^2.$$

Now

$$xu + yv \equiv x^2 + y^2 \equiv 0 \ (\text{mod } m),$$

implying that m divides $xu + yv$, and

$$xv - yu \equiv xy - xy \equiv 0 \ (\text{mod } m),$$

implying that m divides $xv - yu$. Putting $xu + yv = mX$ and $xv - yu = mY$ leads to

$$m^2 rp = m^2 X^2 + m^2 Y^2; \quad \text{that is, } rp = X^2 + Y^2.$$

We can, of course, replace X by $-X$ or Y by $-Y$ if necessary to replace negative integers.

This completes the descent step.

It remains to show that $mp = x^2 + y^2$ has a solution for some m with $1 \leq m < p$. Property (e) of Theorem 2.1 of *Unit 6* provides this solution since it tells us that -1 is a quadratic residue of each prime $p \equiv 1 \ (\text{mod } 4)$. Consequently the congruence $x^2 + 1 \equiv 0 \ (\text{mod } p)$ has a least positive solution x_1 with $0 < x_1 \leq p - 1$. So there exists a positive integer m such that

$$mp = x_1^2 + 1^2,$$

which is exactly as required since

$$m = \frac{x_1^2 + 1^2}{p} \leq \frac{(p-1)^2 + 1}{p} = \frac{p^2 - 2(p-1)}{p} < p.$$

Now if this solution has $m > 1$ then the established descent step guarantees a solution with smaller, positive value of m. We descend through such smaller solutions until we reach one with $m = 1$; that is, a solution of $p = x^2 + y^2$. ∎

The descent step in the above proof may have read like a theoretical argument but, in fact, it contains a construction for getting from an expression of one multiple of p as a sum of two squares to a similar expression for a smaller multiple of p. We can see how the descent step works by looking at a particular example.

Example 4.2

Given that $60^2 + 1^2 = 13 \times 277$, retrace the proof of the descent step in Theorem 4.1 to express the prime 277 as a sum of two squares.

From

$$13 \times 277 = 60^2 + 1^2$$

we have $p = 277$, $m = 13$, $x = 60$ and $y = 1$. So $u \equiv 60 \pmod{13}$ and $v \equiv 1 \pmod{13}$ and as u and v are least absolute residues modulo 13 we have $u = -5$ and $v = 1$. Therefore

$$13r = (-5)^2 + 1^2,$$

so that $r = 2$. Multiplying these two equations together:

$$2 \times 13^2 \times 277 = (60^2 + 1^2)((-5)^2 + 1^2)$$
$$= (-299)^2 + 65^2,$$

and on dividing both sides by 13^2 and removing the minus sign from inside the square,

$$2 \times 277 = 23^2 + 5^2.$$

This completes the first descent step.

As we have not reached 277 itself we descend again. Now $m = 2$, $x = 23$ and $y = 5$ and, replacing x and y by their least absolute residues modulo 2, $u = v = 1$. Then

$$u^2 + v^2 = 1^2 + 1^2 = 2r$$

gives $r = 1$. Multiplying the two equations together

$$2^2 \times 277 = (23^2 + 5^2)(1^2 + 1^2) = 28^2 + 18^2,$$

and on dividing both sides by 2^2,

$$277 = 14^2 + 9^2.$$

We have thus expressed 277 as a sum of two squares. ◆

Knowing that a prime $p = 4k + 1$ can be expressed as $x^2 + y^2$, mathematicians understandably took up the challenge of finding ways of constructing the integers x and y in terms of the prime p. Several such constructions have emerged, but none of them is particularly easy to employ. The first, due to Legendre, showed how to obtain x and y from the continued fraction of $\sqrt{p}$. One disadvantage of this method is that we first have to obtain the continued fraction of $\sqrt{p}$ which can be no mean task when p is large. The next, due to Gauss, is easy to state.

> Let prime $p = 4k + 1$, and let x and y be least absolute residues modulo p satisfying $x \equiv \dfrac{(2k)!}{2(k!)^2} \pmod{p}$ and $y \equiv (2k)!x \pmod{p}$. Then $x^2 + y^2 = p$.

This is very neat, but try using it to express 277 ($k = 69$) as a sum of two squares! We shall not be proving it either.

For practical purposes the method of exhaustion offers as good a method as any for expressing moderately sized primes as a sum of two squares. For instance, look again at the prime 277. First note that if $277 = x^2 + y^2$ then one of x^2 or y^2 is less than $\dfrac{277}{2}$, while the other is greater than $\dfrac{277}{2}$. By virtue of Theorem 4.1 we need only check

$$277 - 1^2, \quad 277 - 2^2, \quad 277 - 3^2, \quad \ldots$$

until we find a square, in the certain knowledge that one will turn up by the time we reach $277 - 12^2$, since $12^2 > \dfrac{277}{2}$. In fact $277 - 9^2 = 14^2$.

This method finds just one solution but, as we shall see in a moment, there is only the one solution.

To represent a composite number as a sum of two squares we can combine the suggested exhaustive approach with use of the important identity.

Problem 4.2 ───────────────────────────────────

Express $5321 = 17 \times 313$ as a sum of two squares.

───

4.2 Sums of two squares, completed

We have seen several examples of integers which can be expressed as a sum of squares in more than one way. This is not true of primes. No prime of the form $4k + 3$ can be expressed as a sum of two squares whilst each prime of the form $4k + 1$ can be expressed as a sum of two squares *in a unique way*. The uniqueness assumes, as we have been doing all along, that by squares we mean squares of non-negative integers, and also that the sum of squares $x^2 + y^2$ is regarded as the same expression as $y^2 + x^2$.

Before we start the proof, note that if $p = x^2 + y^2$, where x and y are positive integers, then each of x and y lies strictly between 0 and $\sqrt{p}$; neither can be 0 for otherwise the prime p would be a square. Moreover $\gcd(x, y) = 1$, as any common divisor of x and y must divide $x^2 + y^2$; that is, it must divide p and it certainly cannot be p itself.

> *Theorem 4.2 Uniqueness of representation*
>
> The expression of a prime of form $4k + 1$ as a sum of two squares is unique except for the order of the two summands.

Proof of Theorem 4.2

Suppose that $p = a^2 + b^2 = c^2 + d^2$, where $a > b > 0$ and $c > d > 0$. We must show that $a = c$ and $b = d$ so that the expression for p as a sum of two squares is unique.

From the two expressions for p we have

$$a^2 d^2 - b^2 c^2 = (p - b^2)d^2 - b^2(p - d^2) = p(d^2 - b^2) \equiv 0 \pmod{p}.$$

That is,

$$(ad - bc)(ad + bc) \equiv 0 \pmod{p}.$$

Now Euclid's Lemma tells us that either p divides $ad - bc$ or p divides $ad + bc$. We shall show that the former must be the case. To that end suppose to the contrary that p divides $ad + bc$. As each of a, b, c and d lies strictly between 0 and $\sqrt{p}$ we have $0 < ad + bc < 2p$. It must therefore be the case that $ad + bc = p$. But then

$$p^2 = (a^2 + b^2)(d^2 + c^2) = (ad + bc)^2 + (ac - bd)^2$$
$$= p^2 + (ac - bd)^2$$

so that $ac - bd = 0$. But since $a > b$ and $c > d$ we have $ac > bd$ and so we have reached the required contradiction.

It follows that p divides $ad - bc$. Again, since each of the four integers lies strictly between 0 and $\sqrt{p}$, we have $-p < ad - bc < p$, and consequently $ad = bc$. From this, a divides bc and, since $\gcd(a, b) = 1$, a divides c. Putting $c = ka$ the equation $ad = bc$ becomes $d = kb$, and then

$$p = c^2 + d^2 = k^2(a^2 + b^2) = k^2 p.$$

This implies that $k = 1$ and, in turn, leads to $a = c$ and $b = d$, exactly as required. ∎

We now know exactly which primes can be expressed as a sum of two squares, namely 2 and any prime of the form $4k + 1$, and we know that any product of numbers expressible as a sum of two squares is itself so expressible. Hence any number which has no prime divisor of the form $4k + 3$ is certainly expressible as a sum of two squares. But are these the only ones? A little experimentation reveals that they are not. For example, $18 = 3^2 + 3^2$ is a sum of two squares which does not adhere to the above prescription because it has a prime divisor of the form $4k + 3$, namely 3 itself. The essential point with 18 is that it is immaterial that 3 cannot be written as a sum of two squares because 3^2 divides 18 and

$$18 = 2 \times 3^2 = (1^2 + 1^2)3^2 = 3^2 + 3^2.$$

In general the identity

$$c^2(a^2 + b^2) = (ca)^2 + (cb)^2$$

shows that we can multiply any sum of squares by *any* square and retain a sum of squares. We can build on this observation in the following way. Suppose, for example, that we want to find one way of expressing $2^7 \times 3^4 \times 5 \times 13^3$ as a sum of squares. By first isolating as large a square term as possible we can write

$$2^7 \times 3^4 \times 5 \times 13^3 = (2^3 \times 3^2 \times 13)^2 \times (2 \times 5 \times 13)$$

and the task will be completed when we express the square-free part, $2 \times 5 \times 13$, as a sum of two squares.

$$(2^3 \times 3^2 \times 13)^2 \times 2 \times 5 \times 13$$
$$= (2^3 \times 3^2 \times 13)^2 \times (1^2 + 1^2)(1^2 + 2^2)(2^2 + 3^2)$$
$$= (2^3 \times 3^2 \times 13)^2 \times (1^2 + 1^2)(8^2 + 1^2)$$
$$= (2^3 \times 3^2 \times 13)^2 \times (9^2 + 7^2)$$
$$= (2^3 \times 3^4 \times 13)^2 + (2^3 \times 3^2 \times 7 \times 13)^2$$

Remember that a square-free integer is one which is not divisible by the square of any prime. Every positive integer is square, square-free or can be expressed as a square multiplied by a square-free integer.

We shall make use of this idea in our proof of the following result which gives a complete classification of which integers are sums of two squares.

Theorem 4.3 Sums of two squares

A positive integer n can be expressed as a sum of two squares if, and only if, each of its prime divisors of the form $4k + 3$, if any, occurs to an even power.

Proof of Theorem 4.3

By pulling out the largest square divisor of n we can write $n = m^2 r$, where r is square-free. The theorem asserts that n is expressible as a sum of two squares if, and only if, r is not divisible by any prime of the form $4k + 3$.

We allow the possibility $r = 1$.

To prove the 'if, and only if,' assertion we have to establish the implications both ways, so we break the proof into two parts.

(a) Suppose that r has no prime divisor of the form $4k + 3$. If $r = 1$ then $n = m^2 + 0^2$ and there is nothing to prove. If $r > 1$ then r is a product of one or more primes each of which is either 2 or of the form $4k + 1$. We have seen that such a product r can be expressed as a sum of two squares, so $n = m^2(a^2 + b^2) = (ma)^2 + (mb)^2$.

(b) Suppose that n can be expressed as a sum of two squares, say

$$n = m^2 r = a^2 + b^2.$$

First, any common divisor of a and b may be cancelled as follows. If $\gcd(a, b) = d$ then we can write $a = a_1 d$, $b = b_1 d$, where $\gcd(a_1, b_1) = 1$, and

$$m^2 r = d^2(a_1^2 + b_1^2).$$

As r is square-free d divides m and so, writing $m_1 = \dfrac{m}{d}$

$$m_1^2 r = a_1^2 + b_1^2.$$

Any prime which divides d occurs with exponent 2, or more, on the right-hand side of this equation and so must divide m.

Our task is to show that r does not have a prime divisor of the form $4k + 3$. Aiming for a contradiction suppose that the prime $p = 4k + 3$ divides r. Then

$$a_1^2 + b_1^2 \equiv 0 \pmod{p}; \quad \text{that is, } a_1^2 \equiv -b_1^2 \pmod{p}.$$

Now if p divides a_1 we would have p divides b_1 thereby contradicting $\gcd(a_1, b_1) = 1$. So $\gcd(a_1, p) = \gcd(b_1, p) = 1$ and FLT can be applied giving

$$a_1^{p-1} \equiv b_1^{p-1} \equiv 1 \pmod{p}.$$

Putting $p = 4k + 3$,

$$1 \equiv a_1^{p-1} \equiv a_1^{4k+2} \equiv (a_1^2)^{2k+1} \equiv (-b_1^2)^{2k+1}$$
$$\equiv (-1)^{2k+1}(b_1^2)^{2k+1} \equiv (-1)b_1^{p-1} \equiv -1 \pmod{p}.$$

This cannot possibly be true for an odd prime p and so we have a contradiction, and the result follows. ∎

Problem 4.3

Which integers in the range 1995 to 2005 inclusive can be expressed as a sum of two squares? For any which can be so expressed find one such representation.

Problem 4.4

Use Theorem 4.3 to show that no positive integer which is congruent modulo 9 to either 3 or 6 can be expressed as a sum of two squares. For each of the other seven congruence classes modulo 9 find the smallest positive integer which can be represented as a sum of two squares and the smallest which cannot.

4.3 Sums of three and four squares

For the sake of completeness we include a little more about the three-square and four-square problems. Finding which integers can, and which cannot, be expressed as a sum of three squares is not difficult but, as so often happens in number theory, finding a proof of an assertion for which there is overwhelming numerical evidence is a different matter. We can readily discover, and prove, which numbers cannot be expressed as a sum of three squares, the difficulty arises in proving that all other numbers can be so expressed.

As with sums of two squares we allow 0^2 so that, for example, $1^2 + 1^2 + 0^2$ is a representation of 2 as a sum of three squares.

Theorem 4.4 Sums of three squares

A positive integer can be expressed as a sum of three squares if, and only if, it is not of the form $4^n(8m + 7)$ for some $n \geq 0$, $m \geq 0$.

Proof of Theorem 4.4

We shall give here only the easier half of the 'if, and only if,' proof, namely that no number of the stated form can be expressed as a sum of three squares.

The squares modulo 8 are 0, 1 and 4, and consequently a sum of three squares can be congruent modulo 8 to any of the values 0, 1, 2, 3, 4, 5 or 6, but not to 7. So no number of the form $8m + 7$ can be a sum of three squares.

Now suppose that for some $n \geq 1$ and $m \geq 0$ we have

$$4^n(8m + 7) = x^2 + y^2 + z^2.$$

As the left-hand side is congruent modulo 4 to 0, and as squares modulo 4 are either 0 or 1, it has to be the case that x, y and z are all even. Putting $x = 2x_1$, $y = 2y_1$ and $z = 2z_1$ we get

$$4^{n-1}(8m + 7) = x_1^2 + y_1^2 + z_1^2.$$

This is similar to the method you met in Problem 3.2

If $n - 1 > 1$ then x_1, y_1 and z_1 are still even and the argument can be repeated:

$$4^{n-2}(8m + 7) = x_2^2 + y_2^2 + z_2^2.$$

In this way we descend through powers of 4 until $8m + 7$ itself is expressed as a sum of three squares. But that is impossible, so the assumption that $4^n(8m + 7)$ can be expressed as a sum of three squares must be false. ∎

Problem 4.5

Which, if any, of the following numbers can be expressed as a sum of three squares? For any which can, find such a representation.

(a) 39   (b) 56   (c) 448   (d) 10!

One main reason for the difficulty in proving the second part of Theorem 4.4, namely that any integer not of the given form can be expressed as a sum of three squares, stems from the fact that there is no equivalent of our important identity. It is not true that if two numbers can each be expressed as a sum of three squares then so too can their product. For example,

$$3 = 1^2 + 1^2 + 1^2 \quad \text{and} \quad 5 = 2^2 + 1^2 + 0^2$$

but the product of these numbers, 15, is not a sum of three squares. Ironically, when we progress to the four-square problem the equivalent of our important identity re-emerges. There is an identity expressing the product of two sums of four squares as a sum of four squares. Here it is.

Important Identity for four squares

$$
\begin{aligned}
(a^2 + b^2 + c^2 + d^2)&(w^2 + x^2 + y^2 + z^2) \\
&= (aw + bx + cy + dz)^2 + (ax - bw + cz - dy)^2 \\
&\quad + (ay - bz - cw + dx)^2 + (az + by - cx - dw)^2
\end{aligned}
$$

With the benefit of this identity the four-square problem essentially comes down to the determination of which *primes* can be written as a sum of four squares.

Theorem 4.5 Lagrange's Four Square Theorem

Every positive integer can be expressed as a sum of four squares.

As indicated above, to prove Lagrange's Theorem it is sufficient to show that each prime can be expressed as a sum of four squares. The even prime $2 = 1^2 + 1^2 + 0^2 + 0^2$ can certainly be expressed this way. For the odd primes there is a very similar argument to our proof of Theorem 4.1 which we could use. Suppose that some multiple mp of the odd prime p can be expressed as a sum of four squares, say

$$mp = a^2 + b^2 + c^2 + d^2, \quad 1 \leq m < p.$$

If $m = 1$ we have the required expression. If not, careful algebra allows us to 'descend' to a smaller multiple which is also a sum of four squares,

$$m_1 p = a_1^2 + b_1^2 + c_1^2 + d_1^2,$$

where $1 \leq m_1 < m$.

The final stage is to show that there really is an appropriate multiple of p which is a sum of four squares, so that from this multiple we can descend in a finite number of steps to p itself being a sum of four squares. The details of the descent step are quite intricate and we shall omit the proof here.

Having established that four squares suffice to represent any non-negative integer, a natural follow-up question is to ask how many cubes, fourth powers, fifth powers, ... are needed. In 1770 Waring proposed the following conjecture.

Theorem 4.6 Waring's Problem

For each integer $k \geq 2$ there exists a positive integer $g(k)$ such that every positive integer can be expressed as a sum of at most $g(k)$ kth powers.

The assertion is that for each k a number $g(k)$ exists, but it carries with it the associated problem of finding values for $g(k)$. The amount of research that has gone into the investigation of various values of $g(k)$ have made

Waring's Problem one of the most actively pursued areas of Number Theory. In 1909 Hilbert proved Waring's original conjecture. For each $k \geq 2$ the number $g(k)$ does exist. But Hilbert's proof was highly theoretical and offered little insight into the values of $g(k)$.

Lagrange's Four Square Theorem confirms that $g(2) = 4$. This was known to Waring, who went on to claim that $g(3) = 9$ and $g(4) = 19$, both of which have now been shown to be true, the latter only very recently. The former is claiming that every positive integer can be expressed as a sum of at most nine cubes. In fact there are just two such integers which require nine cubes:

$$23 = 2^3 + 2^3 + 1^3 + 1^3 + 1^3 + 1^3 + 1^3 + 1^3 + 1^3$$

and

$$239 = 4^3 + 4^3 + 3^3 + 3^3 + 3^3 + 3^3 + 1^3 + 1^3 + 1^3.$$

All integers exceeding 239 can be expressed as a sum of at most eight cubes, and it has been shown that only finitely many of them do require eight cubes, so that from some point onwards seven cubes will suffice.

Back in 1772, J.A. Euler, son of Leonhard, discovered a lower bound for $g(k)$:

$$g(k) \geq \operatorname{int}\left(\left(\frac{3}{2}\right)^k\right) + 2^k - 2.$$

We shall not pause to do so here, but it is not difficult to prove that this inequality holds. What is surprising is that this easily obtained lower bound turns out to give the true value of $g(k)$ for all k so far verified, and very likely for all k. It is now known that $g(k) = \operatorname{int}\left(\left(\frac{3}{2}\right)^k\right) + 2^k - 2$ for all $2 \leq k \leq 200\,000$, and that there are at most a finite number of exceptions after this point.

You might like to check that what the identity gives for $k = 2$, 3 and 4 agrees with what we have already found.

Diversion

The English number theorist G.H. Hardy tells the following story concerning his young protégé Ramanujan: I remember going to visit him in Putney hospital. I had travelled there in taxi cab Number 1729 and I remarked that the number seemed a rather dull one to me; I hoped it was not an unfavourable omen. He replied: 'On the contrary it is a very interesting number; it is the smallest number which can be expressed as the sum of two cubes in two different ways'!

ADDITIONAL EXERCISES

Section 1

1 Find two positive solutions of each of the Diophantine equations
   (a) $x^2 - 14y^2 = 1$ and   (b) $x^2 - 18y^2 = 1$.

2 Find two positive integer solutions of each of the equations

$$x^2 - 17y^2 = \pm 1.$$

3 Show that if $x = x_1$, $y = y_1$ is a solution of $x^2 - ny^2 = -1$ then $x = 2x_1^2 + 1$, $y = 2x_1 y_1$ satisfies $x^2 - ny^2 = 1$. Use this fact to find a solution of $x^2 - 74y^2 = 1$, given that $\sqrt{74} = [8, \langle 1, 1, 1, 1, 16 \rangle]$.

4 (a) Show that the equation $x^2 - ny^2 = -1$ has no solution if
$n \equiv 3 \pmod 4$.

(b) Show that if $x^2 - ny^2 = m$ has solutions, where m and n are
relatively prime, then m is a quadratic residue of each odd prime
divisor of n.

Confirm that the equation $x^2 - 34y^2 = -1$ gives a counter-example
to the converse of this result. Note that $\sqrt{34} = [5, \langle 1, 4, 1, 10 \rangle]$.

5 Find three positive solutions of $x^2 + 2xy - 2y^2 = 1$.

6 The number 48 has the curious property that if 1 is added to it the
result is a square (49), whilst if 1 is added to half of it the result is also
a square (25). Find two more positive integers with this property.

Section 2

1 If (x, y, z) is a primitive Pythagorean triple prove that each of $x + y$
and $x - y$ is congruent modulo 8 to either 1 or 7.

2 Find all Pythagorean triples for which the associated right-angled
triangle has its area numerically equal to its perimeter.

3 Show that if (x, y, z) is a Pythagorean triple then at least one of x or y
is divisible by 3, at least one of x or y is divisible by 4 and at least one
of x, y or z is divisible by 5.

Section 3

1 Suppose that $\sqrt{5} = \dfrac{m}{n}$, where m and n are positive integers. Show that
$\sqrt{5} = \dfrac{5n - 2m}{m - 2n}$. Deduce from this, using the method of infinite descent,
that $\sqrt{5}$ is irrational.

2 Show that it is impossible to find four positive integers which have the
property that the sum of their squares is divisible by twice their
product. Use the the following method.

(a) Consider the Diophantine equation
$$w^2 + x^2 + y^2 + z^2 = 8kwxyz.$$

Show that, in any solution, each of w, x, y and z must be even.
Hint: Consider the two sides of the equation modulo 8.

Then use infinite descent to prove that this equation has no
positive solutions.

(b) Deduce that the Diophantine equation
$$w^2 + x^2 + y^2 + z^2 = 2wxyz$$

has no positive solutions.

3 The Diophantine equation $x^4 - y^4 = 2z^2$ can be shown to have no
solution in which x and y are both odd. Assuming this fact, use the
method of infinite descent to show that this equation has no positive
solution.

4 Show that the Diophantine equation $x^2 + y^2 = x^2y^2$ has no positive
solutions.

Section 4

1   Express each of the numbers 245, 260 and 245×260 as a sum of two squares.

2   If m and n can each be expressed as a sum of two squares and m divides n, is it necessarily true that $\dfrac{n}{m}$ can be expressed as a sum of two squares?

Either prove or give a counter-example, as appropriate.

3   Show that if $p = 4k + 1$ is prime then $2p$ can be written as a sum of two squares in a way which is unique apart from the order of the summands.

4   Find the three smallest integers greater than 1000 which cannot be expressed as a sum of three squares.

5   Show that the number n can be expressed as a sum of three triangular numbers if, and only if, $8n + 3$ can be expressed as a sum of three squares. Hence deduce that every positive integer is a sum of three triangular numbers.

Challenge Problems

1   Prove that if $\dfrac{a}{b}$ is a convergent of $\sqrt{n}$ then $x = a$, $y = b$ is a solution of exactly one of the equations $x^2 - ny^2 = k$, where $|k| < 1 + 2\sqrt{n}$.

2   Let $x = x_k$, $y = y_k$ be the successive solutions of the equation $x^2 - ny^2 = 1$, with $x = x_1$, $y = y_1$ being the fundamental solution. Show that

$$x_{k+1} = 2x_1 x_k - x_{k-1} \quad \text{and} \quad y_{k+1} = 2x_1 y_k - y_{k-1},$$

for all integers ≥ 2.

3   Show that 169 can be expressed as a sum of one, of two, of three and of four *non-zero* squares. Hence show that every integer greater than 169 is a sum of five *non-zero* squares.

Which positive integers cannot be expressed as a sum of five non-zero squares?

4   Prove that the Diophantine equation $x^2 + y^2 = w^3 + z^3$ does not have a positive solution in the case where y is odd and has no prime divisor of the form $4k + 3$ and z is congruent modulo 4 to 2.

Hence show that $x^2 = y^3 + 7$ has no solution.

SOLUTIONS TO THE PROBLEMS

Solution 1.1

(a) The convergents of of $\sqrt{3} = [1, \langle 1, 2 \rangle]$ begin as shown in the following table.

k	1	2	3	4	5	6	7	8	9	10
p_k	1	2	5	7	19	26	71	97	265	362
a_k	1	1	2	1	2	1	2	1	2	1
q_k	1	1	3	4	11	15	41	56	153	209
$p_k^2 - 3q_k^2$	-2	1	-2	1	-2	1	-2	1	-2	1

It appears that every even convergent gives a solution of this Pell's equation.

(b) For $\sqrt{10} = [3, \langle 6 \rangle]$ we have:

k	1	2	3	4	5	6
p_k	3	19	117	721	4443	27379
a_k	3	6	6	6	6	6
q_k	1	6	37	228	1405	8658
$p_k^2 - 10q_k^2$	-1	1	-1	1	-1	1

Once again the even convergents appear to give solutions of Pell's equation, the first three of which give:

$$19^2 - 10 \times 6^2 = 1;$$
$$721^2 - 10 \times 228^2 = 1;$$
$$27379^2 - 10 \times 8658^2 = 1.$$

Solution 1.2

The convergents of $\sqrt{11} = [3, \langle 3, 6 \rangle]$ are

$$\frac{3}{1}, \frac{10}{3}, \frac{63}{19}, \frac{199}{60}, \frac{1257}{379}, \frac{3970}{1197}, \cdots$$

As $\sqrt{11}$ has a cycle of even length 2, the even convergents will all give rise to solutions of this Pell's equation. The first three solutions are therefore given by the convergents C_2, C_4 and C_6 as:

$$x = 10, \ y = 3, \quad 10^2 - 11 \times 3^2 = 1;$$
$$x = 199, \ y = 60, \quad 199^2 - 11 \times 60^2 = 1;$$
$$x = 3970, \ y = 1197, \quad 3970^2 - 11 \times 1197^2 = 1.$$

Solution 1.3

As $\sqrt{13} = [3, \langle 1, 1, 1, 1, 6 \rangle]$ has cycle of odd length 5, the convergent $\dfrac{p_5}{q_5}$ satisfies $p_5^2 - 13q_5^2 = -1$ and the convergent $\dfrac{p_{10}}{q_{10}}$ gives the smallest solution of $x^2 - 13y^2 = 1$.

The convergents of $\sqrt{13}$ begin

$$\frac{3}{1}, \frac{4}{1}, \frac{7}{2}, \frac{11}{3}, \frac{18}{5}, \frac{119}{33}, \frac{137}{38}, \frac{256}{71}, \frac{393}{109}, \frac{649}{180},$$

and we can confirm that:

$$18^2 - 13 \times 5^2 = -1,$$

giving $x = 18$, $y = 5$ as a solution of $x^2 - 13y^2 = -1$;

$$649^2 - 13 \times 180^2 = 1,$$

giving $x = 649$, $y = 180$ as a solution of $x^2 - 13y^2 = +1$.

Solution 1.4

The first convergent, $C_1 = \dfrac{4}{1}$ gives the smallest solution $x_1 = 4$, $y_1 = 1$ of $x^2 - 17y^2 = -1$.

As

$$(4 + \sqrt{17})^2 = 33 + 8\sqrt{17},$$

we have $x = 33$, $y = 8$ as a solution of $x^2 - 17y^2 = 1$. As

$$(4 + \sqrt{17})^4 = (33 + 8\sqrt{17})^2 = 2177 + 528\sqrt{17},$$

we have $x = 2177$, $y = 528$ as the next solution of $x^2 - 17y^2 = 1$.

Solution 1.5

As $\sqrt{2} = [1, \langle 2 \rangle]$, which has cycle of length 1, every even convergent gives a solution of the Pell's equation. The first ten convergents of $\sqrt{2}$ are as follows.

$$\frac{1}{1}, \frac{3}{2}, \frac{7}{5}, \frac{17}{12}, \frac{41}{29}, \frac{99}{70}, \frac{239}{169}, \frac{577}{408}, \frac{1393}{985}, \frac{3363}{2378}.$$

For each convergent $\dfrac{p_{2k}}{q_{2k}}$, the required triangular square is $\left(\dfrac{q_{2k}}{2}\right)^2$. The first five are:

$$\left(\frac{2}{2}\right)^2 = 1, \quad \left(\frac{12}{2}\right)^2 = 36, \quad \left(\frac{70}{2}\right)^2 = 1225, \quad \left(\frac{408}{2}\right)^2 = 41\,616, \quad \left(\frac{2378}{2}\right)^2 = 1\,413\,721.$$

Solution 2.1

(a) $(10, 8, 6)$ is not a Pythagorean triple; the hypoteneuse must be the largest of the three values and, by convention, must be the last member of the triple.

(b) $(12, 9, 15)$ is a Pythagorean triple, but is not primitive since it is the $(4, 3, 5)$ triple scaled by a multiple of 3.

(c) $(6, 7, 8)$ is not a Pythagorean triple : $6^2 + 7^2 \neq 8^2$.

(d) $(5, 12, 13)$ is a primitive Pythagorean triple: $5^2 + 12^2 = 169 = 13^2$ and the three integers are relatively prime in pairs.

(e) $(24, 33, 41)$ is not a Pythagorean triple since 24 and 33 have a common divisor, namely 3, which does not divide the third number.

Solution 2.2

We seek relatively prime positive integers m and n, of opposite parity and with $m > n$, such that $2mn = 72$. There are two such pairs of integers:

$$m = 36, \ n = 1, \ \text{giving the triple } (72, 1295, 1297);$$
$$m = 9, \ n = 4, \ \text{giving the triple } (72, 65, 97).$$

The pairs $m = 18$, $n = 2$ and $m = 12$, $n = 3$ are excluded because they are not relatively prime.

Solution 2.3

One possible strategy is to look at each divisor of 30 in turn and to determine all primitive triples containing that divisor. But notice that, since 30 is not a multiple of 4, no even divisor of 30 can occur in a primitive Pythagorean triple. So we need look only at the odd divisors of 30. Any primitive triple containing one of 3, 5 or 15 may be scaled appropriately to give a Pythagorean triple containing 30.

Remember that the even member in a primitive triple is divisible by 4.

We therefore seek all pairs of relatively prime integers m and n of opposite parity and with $m > n$ such that either $m^2 + n^2$ or $m^2 - n^2$ is equal to one of 3, 5 or 15. It turns out that there are just five such pairs, as presented below.

Divisor	m	n	Primitive triple	Scaling	Triple with 30
15	4	1	$(8, 15, 17)$	2	$(16, 30, 34)$
15	8	7	$(112, 15, 113)$	2	$(224, 30, 226)$
5	3	2	$(12, 5, 13)$	6	$(72, 30, 78)$
5	2	1	$(4, 3, 5)$	6	$(24, 18, 30)$
3	2	1	$(4, 3, 5)$	10	$(40, 30, 50)$

As $m^2 - n^2 = (m - n)(m + n)$ we need only consider m and n for which $m + n \leq 15$.

Solution 2.4

Suppose that $st = u^2$, where s and t are relatively prime. Let s and t have prime decompositions $s = p_1^{k_1} p_2^{k_2} \ldots p_i^{k_i}$, $t = p_{i+1}^{k_{i+1}} p_{i+2}^{k_{i+2}} \ldots p_j^{k_j}$. As $\gcd(s, t) = 1$ the listed primes are all distinct and so st has prime decomposition

$$st = p_1^{k_1} \ldots p_i^{k_i} p_{i+1}^{k_{i+1}} \ldots p_j^{k_j}.$$

The primes are not necessarily listed in ascending order here.

As st is a square, each exponent in this expression is even. Hence each exponent in s, and each exponent in t, is even and so these two are squares as well.

Solution 2.5

Suppose that $(x, x + d, x + 2d)$ is a Pythagorean triple. Then

$$x^2 + (x + d)^2 = (x + 2d)^2$$

To fit our convention the triple would actually be $(x + d, x, x + 2d)$.

giving

$$x^2 - 2xd - 3d^2 = 0.$$

This factorizes as

$$(x - 3d)(x + d) = 0.$$

Hence $x = 3d$ or $x = -d$. As the latter cannot lead to positive solutions we are left with $x = 3d$, which gives the triple $(3d, 4d, 5d)$. The only Pythagorean triples which have the three numbers in arithmetic progression are the multiples of $(3, 4, 5)$.

Solution 2.6

Suppose that $(m, m + 1, n)$ is a Pythagorean triple, so $m^2 + (m + 1)^2 = n^2$. Then

$$T_{2m}^2 + T_{2m+1}^2 = \left(\frac{2m(2m + 1)}{2}\right)^2 + \left(\frac{(2m + 1)(2m + 2)}{2}\right)^2$$
$$= (2m + 1)^2(m^2 + (m + 1)^2) = ((2m + 1)n)^2$$

which shows that $(T_{2m}, T_{2m+1}, 2mn + n)$ is a Pythagorean triple. Knowing that there are infinitely many primitive Pythagorean triples $(m, m + 1, n)$ we conclude that there are infinitely many whose shorter sides are consecutive triangular numbers.

Solution 3.1

Suppose that $x_1^4 + 2y_1^4 = 4z_1^4$, where x_1, y_1 and z_1 are positive integers. As the prime 2 divides two of the terms it must divide the third, namely x_1^4. So 2 divides x_1 and we can write $x_1 = 2x_2$ for some positive integer x_2. Substituting for x_1 gives

$$16x_2^4 + 2y_1^4 = 4z_1^4; \quad \text{that is, } 8x_2^4 + y_1^4 = 2z_1^4.$$

Repeating the argument, we can write $y_1 = 2y_2$ for some positive integer y_2 and obtain

$$8x_2^4 + 16y_2^4 = 2z_1^4; \quad \text{that is, } 4x_2^4 + 8y_2^4 = z_1^4.$$

We can now write $z_1 = 2z_2$ for some positive integer z_2 and

$$4x_2^4 + 8y_2^4 = 16z_2^4; \quad \text{that is, } x_2^4 + 2y_2^4 = 4z_2^4.$$

At this point we have reached a second positive solution, $x = x_2$, $y = y_2$, $z = z_2$ of the original equation with $z_2 < z_1$. The descent step is therefore complete and the required contradiction established.

Thus $x^4 + 2y^4 = 4z^4$ has no positive solutions.

Solution 3.2

(a) As any square is congruent modulo 4 to 0 (if it is even) or to 1 (if it is odd) it follows that $x^2 + y^2 + z^2 \equiv 0 \pmod 4$ can only occur when x, y and z are all even.

(b) Suppose that $x_1^2 + y_1^2 + z_1^2 = 2x_1y_1z_1$. If x_1, y_1 and z_1 are all odd then $x_1^2 + y_1^2 + z_1^2 \equiv 3 \pmod 4$ while the right-hand side, $2x_1y_1z_1$, is even. Hence at least one of x_1, y_1 and z_1 must be even. But then $x_1^2 + y_1^2 + z_1^2 = 2x_1y_1z_1 \equiv 0 \pmod 4$ and, as we have just seen, x_1, y_1 and z_1 must all be even.

Writing $x_1 = 2x_2$, $y_1 = 2y_2$ and $z_1 = 2z_2$ we have

$$(2x_2)^2 + (2y_2)^2 + (2z_2)^2 = 2(2x_2)(2y_2)(2z_2),$$

which simplifies to

$$x_2^2 + y_2^2 + z_2^2 = 4x_2y_2z_2.$$

As the right-hand side is congruent modulo 4 to 0, the first part of the question once again gives that x_2, y_2, and z_2 are even positive integers.

(c) Continuing from part (b), if we write $x_2 = 2x_3$, $y_2 = 2y_3$, $z_2 = 2z_3$ we obtain

$$x_3^2 + y_3^2 + z_3^2 = 8x_3y_3z_3$$

with x_3, y_3 and z_3 being even positive integers. We can continue forever in this way, halving the x, y and z values and yet retaining even positive integers; for from

$$x_n^2 + y_n^2 + z_n^2 = 2^n x_n y_n z_n$$

we note that x_n, y_n and z_n are even integers and

$$\left(\frac{x_n}{2}\right)^2 + \left(\frac{y_n}{2}\right)^2 + \left(\frac{z_n}{2}\right)^2 = 2^{n+1} \left(\frac{x_n}{2}\right) \left(\frac{y_n}{2}\right) \left(\frac{z_n}{2}\right).$$

The same argument now confirms that $\frac{x_n}{2}$, $\frac{y_n}{2}$ and $\frac{z_n}{2}$, are still even positive integers. As the sequence $z_1, z_2, z_3, \ldots$ of positive integers cannot decrease indefinitely, the method of infinite descent shows that the supposed positive solution of the original equation cannot exist.

Solution 3.3

From $x^2 + y^2 = z^2$ and $xy = 2n^2$ we have

$$(x + y)^2 = x^2 + y^2 + 2xy = z^2 + (2n)^2$$

and

$$(x - y)^2 = x^2 + y^2 - 2xy = z^2 - (2n)^2.$$

Multiplying these two equations together gives

$$(x^2 - y^2)^2 = z^4 - (2n)^4.$$

This contradicts the Corollary to Theorem 3.2. We conclude that no Pythagorean triangle can have an area which is a square.

Note $x \neq y$ in any Pythagorean triple and $x^2 - y^2$ can be replaced by $y^2 - x^2$ if the former is negative.

Solution 4.1

$$
\begin{aligned}
325 = 5^2 \times 13 &= (2^2 + 1^2)(2^2 + 1^2)(3^2 + 2^2) \\
&= (2^2 + 1^2)(8^2 + 1^2) = 17^2 + 6^2 \\
&= (1^2 + 2^2)(8^2 + 1^2) = 10^2 + 15^2 \\
&= (2^2 + 1^2)(2^2 + 3^2)(2^2 + 1^2) = (7^2 + 4^2)(2^2 + 1^2) = 18^2 + 1^2
\end{aligned}
$$

Solution 4.2

As $313 \equiv 1 \pmod 4$ we know that 313 can be expressed as a sum of two squares. Searching the sequence

$$313 - 1^2, \quad 313 - 2^2, \quad 313 - 3^2, \quad \ldots$$

for the first square we find $313^2 - 12^2 = 13^2$. Therefore

$$5321 = 17 \times 313 = (1^2 + 4^2)(12^2 + 13^2) = 64^2 + 35^2.$$

In fact there is just one alternative solution which can be reached by writing 313 as $13^2 + 12^2$:

$$5321 = (1^2 + 4^2)(13^2 + 12^2) = 61^2 + 40^2.$$

The expression for a prime is unique, but it may not be so for a composite integer.

Solution 4.3

$1995 = 3 \times 5 \times 7 \times 19$	not expressible since $3 \equiv 3 \pmod 4$
$1996 = 2^2 \times 499$	not expressible since $499 \equiv 3 \pmod 4$
1997 is a prime	$1997 = 34^2 + 29^2$
$1998 = 2 \times 3^3 \times 37$	not expressible since 3 occurs with odd exponent
1999 is a prime	not expressible since $1999 \equiv 3 \pmod 4$
$2000 = 2^4 \times 5^3$	$2000 = 40^2 + 20^2$
$2001 = 3 \times 23 \times 29$	not expressible since $3 \equiv 3 \pmod 4$
$2002 = 2 \times 7 \times 11 \times 13$	not expressible since $7 \equiv 3 \pmod 4$
2003 is a prime	not expressible since $2003 \equiv 3 \pmod 4$
$2004 = 2^2 \times 3 \times 167$	not expressible since $3 \equiv 3 \pmod 4$
$2005 = 5 \times 401$	$2005 = 41^2 + 18^2$

Solution 4.4

If $n \equiv 3 \pmod 9$ then $n = 9k + 3 = 3(3k + 1)$ for some integer k. Now $\gcd(3, 3k + 1) = 1$ and so, in the prime decomposition of n, 3 occurs with exponent 1. Theorem 4.3 confirms that n cannot be expressed as a sum of two squares.

Similarly, if $n = 9k + 6 = 3(3k + 2)$ then n is divisible by 3 but not by 3^2 and so is not expressible as a sum of two squares.

Taking each of the other congruence classes modulo 9 in turn:

Residue class modulo 9	Smallest sum of two squares	Smallest not sum of two squares
0	$9 = 3^2 + 0^2$	$27 = 3^3$
1	$1 = 1^2 + 0^2$	19
2	$2 = 1^2 + 1^2$	11
4	$4 = 2^2 + 0^2$	$22 = 2 \times 11$
5	$5 = 2^2 + 1^2$	$14 = 2 \times 7$
7	$16 = 4^2 + 0^2$	7
8	$8 = 2^2 + 2^2$	$35 = 5 \times 7$

Each of the primes 3, 7, 11 and 19 is of the form $4k + 3$.

Solution 4.5

(a) $39 \equiv 7 \pmod 8$ and so cannot be expressed as a sum of three squares.

(b) $56 = 4 \times 14$ is not of form $4^n(8m + 7)$ and so can be written as a sum of three squares:

$$56 = 6^2 + 4^2 + 2^2.$$

(c) $448 = 4^3 \times 7$ cannot be expressed as a sum of two squares.

(d) $10! = 2^8 \times 3^4 \times 5^2 \times 7 = 4^4(3^4 \times 5^2 \times 7)$. Now as $3^2 \equiv 5^2 \equiv 1 \pmod 8$ it follows that $3^4 \times 5^2 \times 7 \equiv 7 \pmod 8$ and so $10!$ cannot be expressed as a sum of three squares.

SOLUTIONS TO ADDITIONAL EXERCISES

Section 1

1   (a) $\sqrt{14} = [3, \langle 1, 2, 1, 6 \rangle]$ and has convergents

$$\frac{3}{1}, \frac{4}{1}, \frac{11}{3}, \mathbf{\frac{15}{4}}, \frac{101}{27}, \frac{116}{31}, \frac{333}{89}, \mathbf{\frac{449}{120}}, \cdots$$

The convergents, shown in bold-faced type, which come immediately before the end of the cycle give the required solutions.

$$15^2 - 14 \times 4^2 = 1, \quad \text{so } x = 15, \ y = 4$$

and

$$449^2 - 14 \times 120^2 = 1, \quad \text{so } x = 449, \ y = 120.$$

   (b) $\sqrt{18} = [4, \langle 4, 8 \rangle]$ and has convergents

$$\frac{4}{1}, \mathbf{\frac{17}{4}}, \frac{140}{33}, \mathbf{\frac{577}{136}}, \cdots$$

The convergents, shown in bold-faced type, give the two required solutions.

$$17^2 - 18 \times 4^2 = 1, \quad \text{so } x = 17, \ y = 4$$

and

$$577^2 - 18 \times 136^2 = 1, \quad \text{so } x = 577, \ y = 136.$$

2 $\sqrt{17} = [4, \langle 8 \rangle]$ and has convergents

$$\frac{4}{1}, \frac{33}{8}, \frac{268}{65}, \frac{2177}{528}, \ldots$$

As the cycle has length 1 the convergents alternately give solutions of $x^2 - 17y^2 = -1$ and $x^2 - 17y^2 = 1$.

$$4^2 - 17 \times 1^2 = -1, \quad 33^2 - 17 \times 8^2 = 1,$$
$$268^2 - 17 \times 65^2 = -1, \quad 2177^2 - 17 \times 528^2 = 1.$$

So the solutions of $x^2 - 17y^2 = -1$ are $x = 4$, $y = 1$ and $x = 268$, $y = 65$, and the solutions of $x^2 - 17y^2 = +1$ are $x = 33$, $y = 8$ and $x = 2177$, $y = 528$.

3
$$x^2 - ny^2 = (2x_1^2 + 1)^2 - n(2x_1 y_1)^2 = 4x_1^4 + 4x_1^2 + 1 - 4nx_1^2 y_1^2$$
$$= 4x_1^2(x_1^2 - ny_1^2) + 4x_1^2 + 1$$
$$= 4x_1^2(-1) + 4x_1^2 + 1 = 1.$$

The first solution of $x^2 - 74y^2 = -1$ will arise from the convergent $[8, 1, 1, 1, 1]$. Working systematically through the convergents:

$$\frac{8}{1}, \frac{9}{1}, \frac{17}{2}, \frac{26}{3}, \mathbf{\frac{43}{5}} = [8, 1, 1, 1, 1].$$

The solution of $x^2 - 74y^2 = 1$, given by the first part of the question is then

$$x = 2 \times 43^2 + 1 = 3699, \quad y = 2 \times 43 \times 5 = 430.$$

Check: $43^2 - 74 \times 5^2 = -1$.

Check: $3699^2 - 74 \times 430^2 = 1$.

In fact this is the smallest positive solution (given by the numerator and denominator of the convergent $[8, 1, 1, 1, 1, 16, 1, 1, 1, 1]$).

4 (a) A square is congruent modulo 4 to either 0 or 1. Hence, if $n \equiv 3 \pmod 4$,

$$x^2 - ny^2 \equiv \{0 \text{ or } 1\} - \{0 \text{ or } 3\} \equiv \{0, 1 \text{ or } 2\} \pmod 4.$$

In this case $x^2 - ny^2 = -1$ is impossible.

(b) Let p be an odd prime divisor of n. Then

$$x^2 \equiv m \pmod p$$

and so m is a quadratic residue of p.

For the equation $x^2 - 34y^2 = -1$, we note that 17 is the only odd prime divisor of 34 and -1 is a quadratic residue of 17 (by Euler's Criterion). But the equation is known to have no solutions because $\sqrt{34} = [5, \langle 1, 4, 1, 10 \rangle]$ has a cycle of even length.

5 A little algebra reveals the connection with Pell's equation:

$$x^2 + 2xy - 2y^2 = 1 \text{ can be written as } (x + y)^2 - 3y^2 = 1.$$

Substituting $z = x + y$ we obtain

$$z^2 - 3y^2 = 1$$

whose solutions are found from the convergents of $\sqrt{3} = [1, \langle 1, 2 \rangle]$ which are

$$\frac{1}{1}, \mathbf{\frac{2}{1}}, \frac{5}{3}, \mathbf{\frac{7}{4}}, \frac{19}{11}, \mathbf{\frac{26}{15}}, \ldots$$

The three convergents shown in bold give the three required solutions:

$$z = 2, \ y = 1 \text{ gives } x = 1, \ y = 1;$$
$$z = 7, \ y = 4 \text{ gives } x = 3, \ y = 4;$$
$$z = 26, \ y = 15 \text{ gives } x = 11, \ y = 15.$$

6   Let the number be x. The properties it has to possess are

$$x + 1 = y^2 \quad \text{and} \quad \frac{x}{2} + 1 = z^2.$$

Eliminating x from this pair of equations produces $y^2 - 2z^2 = -1$. As the continued fraction $\sqrt{2} = [1, \langle 2 \rangle]$ has a cycle of length 1, every odd convergent yields a solution of this equation. The convergents begin

$$\frac{\mathbf{1}}{\mathbf{1}}, \frac{3}{2}, \frac{\mathbf{7}}{\mathbf{5}}, \frac{17}{12}, \frac{\mathbf{41}}{\mathbf{29}}, \frac{99}{70}, \frac{\mathbf{239}}{\mathbf{169}},$$

giving the required solutions:

$y = 1$, $z = 1$ and hence $x = 0$ (which we discount as it is not positive);

$y = 7$, $z = 5$ and hence $x = 48$ (the one given);

$y = 41$, $z = 29$ and hence $x = 1680$;

$y = 239$, $z = 169$ and hence $x = 57\,120$.

Section 2

1   If (x, y, z) is a primitive Pythagorean triple with x as the even member then $x = 2mn$ and $y = m^2 - n^2$ for relatively prime integers m and n of opposite parity. Then

$$x + y = 2mn + m^2 - n^2 = (m + n)^2 - 2n^2.$$

As $m + n$ is odd $(m + n)^2 \equiv 1 \pmod 8$, and $2n^2 \equiv 0$ or $2 \pmod 8$ depending on whether n is even or odd respectively.

Hence $x + y \equiv 1$ or $7 \pmod 8$.

Similarly,

$$x - y = 2mn - m^2 + n^2 = -(m - n)^2 + 2n^2 \equiv 1 \text{ or } 7 \pmod 8.$$

2   For the triple (x, y, z) the area of the associated triangle is $\dfrac{xy}{2}$ and the perimeter is $x + y + z$. So we have the two equations

$$x^2 + y^2 = z^2 \quad \text{and} \quad \frac{xy}{2} = x + y + z$$

to solve simultaneously. Eliminating z

$$x^2 + y^2 = \left(\frac{xy}{2} - x - y\right)^2 = \frac{x^2 y^2}{4} + x^2 + y^2 - x^2 y - xy^2 + 2xy,$$

which, after cancelling $x^2 + y^2$, multiplying through by 4 and dividing throughout by xy, reduces to

$$xy - 4x - 4y + 8 = 0$$

and hence to

$$(x - 4)(y - 4) = 8.$$

For positive x and y, the product on the left must be 8×1 or 4×2, there are just two solutions (ignoring those with x and y interchanged), namely $(12, 5, 13)$ and $(8, 6, 10)$.

3   If neither x nor y is divisible by 3 then $x^2 + y^2 \equiv 2 \pmod 3$ which cannot be a square.

Theorem 2.1 showed that the even member in any primitive Pythagorean triple is divisible by 4; in any multiple of this triple it will still be a multiple of 4.

If neither x nor y is divisible by 5 then

$$x^2 + y^2 \equiv \{1 \text{ or } 4\} + \{1 \text{ or } 4\} \equiv \{0, 2 \text{ or } 3\} \pmod 5.$$

As 2 and 3 are not quadratic residues of 5 we must have $x^2 + y^2 \equiv 0 \pmod 5$ which means that z is divisible by 5.

Hence at least one of x, y or z is divisible by 5.

Section 3

1 If $m = \sqrt{5}n$ then

$$\frac{5n - 2m}{m - 2n} = \frac{5 - 2\sqrt{5}}{\sqrt{5} - 2} = \sqrt{5}.$$

Note that $m - 2n = (\sqrt{5} - 2)n$ and since $0 < \sqrt{5} - 2 < 1$ we have $0 < m - 2n < n$. Hence the assumption that $\sqrt{5}$ can be expressed as the quotient of positive integers $\dfrac{m}{n}$ leads to an expression for $\sqrt{5}$ as the quotient of positive integers with a smaller denominator. This gives us our descent step. As we cannot descend forever through positive integers this is a contradiction. Hence $\sqrt{5}$ is irrational.

2 (a) As the right-hand side of the equation

$$w^2 + x^2 + y^2 + z^2 = 8kwxyz$$

is even, then either none, two or all four of the variables w, x, y and z must be odd. Recalling that the square of any odd integer is congruent modulo 8 to 1, whilst the square of an even integer is congruent modulo 8 to either 0 or 4,

$$w^2 + x^2 + y^2 + z^2 \equiv 4 \pmod 8, \text{ when all four are odd}$$

and

$$w^2 + x^2 + y^2 + z^2 \equiv \{2 \text{ or } 6\} \pmod 8, \text{ when two are odd.}$$

As the right-hand side of the equation is congruent modulo 8 to 0, the only possibility, therefore, is that all four of w, x, y and z are even.

Now we can write $w = 2w_1$, $x = 2x_1$, $y = 2y_1$ and $z = 2z_1$ and the equation becomes

$$w_1^2 + x_1^2 + y_1^2 + z_1^2 = 32kw_1x_1y_1z_1 = 8k_1w_1x_1y_1z_1,$$

for $k_1 = 4k$. This is another positive solution of the original equation (with the variables w, x, y, z and k) in which the four values w, x, y and z have each been halved. We cannot descend for ever through positive integer values in this way, and so the assumption of a solution is contradicted.

The fifth variable, k, has increased in value, but this does not affect the descent of the other four positive integers. In fact we need only consider one variable, for example, z.

(b) By very similar reasoning to that above, in any solution of

$$w^2 + x^2 + y^2 + z^2 = 2wxyz$$

w, x, y and z must all be even. Putting $w = 2w_1$, $x = 2x_1$, $y = 2y_1$ and $z = 2z_1$ the equation becomes

$$w_1^2 + x_1^2 + y_1^2 + z_1^2 = 8w_1x_1y_1z_1,$$

which we have seen has no solutions.

If all four are odd the left-hand side is divisible by 4 but the right is not. If two are odd and two even the right-hand side is divisible by 4 but the left is not.

3 First note that, as $x^4 - y^4$ has to be even, in any solution x and y have the same parity. As we are given that there is no solution with both x and y odd we are left with the task of showing there is no solution in which both are even. So, suppose to the contrary that there is a positive solution in which x and y are both even. Putting $x = 2x_1$ and $y = 2y_1$ the equation becomes $16(x_1^4 - y_1^4) = 2z^2$. As the left-hand side of this equation is divisible by 16 we must have that z is a multiple of 4, and putting $z = 4z_1$ we get $x_1^4 - y_1^4 = 2z_1^2$. This completes the descent step as $0 < z_1 < z$. The assumption that there exists a positive solution has lead to the existence of a smaller positive one; an impossible situation. Hence the equation has no positive solution.

> The smaller solution must still have x_1 and y_1 even as the alternative has been excluded.

4 Suppose that $x^2 + y^2 = x^2 y^2$. Then, rearranging the equation,

$$(x^2 - 1)(y^2 - 1) = 1.$$

For the product of two integers to give 1, either each bracket on the left is equal to 1 or each is equal to -1. Hence $x = y = \sqrt{2}$ or $x = y = 0$, neither of which gives a positive integer solution.

Section 4

1
$$245 = 5 \times 7^2 = (1^2 + 2^2)7^2 = 7^2 + 14^2$$
$$260 = 2^2 \times 5 \times 13 = 2^2(1^2 + 2^2)(2^2 + 3^2) = 2^2(8^2 + 1^2) = 16^2 + 2^2$$
$$245 \times 260 = (7^2 + 14^2)(16^2 + 2^2) = 140^2 + 210^2$$

2 The result is true. If m and n are each sums of two squares then, for each prime p of the form $4k + 3$ which divides n, the exponent of p in n is even and the exponent of p in m is even (possibly 0). Now the exponent of p in $\dfrac{n}{m}$ is the difference of these even numbers, and hence even. The result follows from Theorem 4.3.

3 If $p = 4k + 1$ then p can be expressed as a sum of two squares, say $p = a^2 + b^2$. But then

$$2p = (1^2 + 1^2)(a^2 + b^2) = (a + b)^2 + (a - b)^2$$

and so $2p$ can also be expressed as a sum of two squares.

For uniqueness, suppose that $2p = x^2 + y^2$. Then $x^2 + y^2 \equiv 2 \pmod 4$ which means that each of x and y is odd. In that case

$$2p = x^2 + y^2 = 2\left(\left(\frac{x+y}{2}\right)^2 + \left(\frac{x-y}{2}\right)^2\right)$$

which gives the expression

$$p = \left(\frac{x+y}{2}\right)^2 + \left(\frac{x-y}{2}\right)^2$$

for p as a sum of two squares. But we know by Theorem 4.2 that this expression is unique. Therefore $\dfrac{x+y}{2}$ and $\dfrac{x-y}{2}$, and in consequence x and y, are determined uniquely.

4 We seek the three smallest integers exceeding 1000 which can be written in the form $4^n(8m + 7)$. Any integer which is congruent to 7 modulo 8 is of this form and the only other candidates are multiples of 4. Starting at 1000, we examine numbers of either of these forms, as follows:

$1004 = 4 \times 251$ and $251 \equiv 3 \pmod 8$;

$1007 \equiv 7 \pmod 8$, so 1007 is not a sum of three squares;

$1008 = 4^2 \times 63$ and $63 \equiv 7 \pmod 8$, so 1008 is not a sum of three squares;

$1012 = 4 \times 253$ and $253 \equiv 5 \pmod 8$;

$1015 \equiv 7 \pmod 8$, so 1015 is not a sum of three squares.

5 The triangular numbers are $T_r = \dfrac{r(r + 1)}{2}$, for $r \geq 1$. So if $n = T_r + T_s + T_t$ then

$$n = \frac{r(r + 1)}{2} + \frac{s(s + 1)}{2} + \frac{t(t + 1)}{2}.$$

So

$$8n + 3 = 4r(r + 1) + 4s(s + 1) + 4t(t + 1) + 3$$
$$= (2r + 1)^2 + (2s + 1)^2 + (2t + 1)^2$$

which confirms that $8n + 3$ is a sum of three squares.

The converse also holds. Starting with this expression for $8n + 3$ as the sum of three squares (which must all be odd) we reverse the algebraic steps and recover the expression for n as a sum of three triangular numbers.

As no integer of the form $8n + 3$ is also of the form $4^r(8s + 7)$, Theorem 4.4 has the corollary that every positive integer is a sum of three triangular numbers.

INDEX